Asian Aerospace and Astronautics Conference (AAAC 2023)

Journal of Physics: Conference Series Volume 2633

Wuhan, China
15 - 17 September 2023

ISBN: 978-1-7138-8217-6
ISSN: 1742-6588

Printed from e-media with permission by:

Curran Associates, Inc.
57 Morehouse Lane
Red Hook, NY 12571

Some format issues inherent in the e-media version may also appear in this print version.

Institute of Physics
Dirac House, Temple Back
Bristol BS1 6BE UK

Phone: 44 1 17 929 7481
Fax: 44 1 17 920 0979

techtracking@iop.org

Additional copies of this publication are available from:

Curran Associates, Inc.
57 Morehouse Lane
Red Hook, NY 12571 USA
Phone: 845-758-0400
Fax: 845-758-2633
Email: curran@proceedings.com
Web: www.proceedings.com

TABLE OF CONTENTS

Journal of Physics: Conference Series **2633** (2023) 011001 doi:10.1088/1742-6596/2633/1/011001

Preface

The Asian Aerospace and Astronautics Conference (AAAC 2023) will take place in Wuhan, Hubei, China from September 15-17, 2023, where participants will gain detailed insights into the state of the art of aerospace and astronautics, and enjoy the exchange with other enthusiasts from all over the world who are interested in this highly relevant and constantly growing area. The conference is sponsored by Science and Engineering Institute (SCIEI), Huazhong University of Science and Technology, co-sponsored by The Hong Kong Polytechnic University, Nanjing University of Science and Technology, York University, etc.

This international conference was over three-days and provided a forum for discussion of aerospace and astronautics. Submitted papers were received and sent for preliminary and peer review conducted by the technical committees. Professional comments are given according to the aspects of originality, innovation, applicability, technical merit, organizing and writing, relevance to conference. After several rounds of review procedure, some excellent papers have been received to get published in the conference proceedings. The proceedings contains a collection of research papers, organized with 6 chapters: aircraft structural design and wing aerodynamic analysis, automation control and visualization technology for flight systems, system design and safety analysis in aerospace engineering, aviation system and engine performance simulation and evaluation, new propulsion technology based on combustion mode, engineering material mechanics analysis and performance simulation and so on.

The conference invited four internationally recognised experts in their fields of aerospace and astronautics to deliver keynote and invite speeches, including Prof. Zheng Hong Zhu, York University, Canada; Prof. Chih-Yung Wen, The Hong Kong Polytechnic University, Hong Kong, China; Prof. Yan Wang, Nanjing University of Aeronautics and Astronautics, China; Prof. Lei Shi, Northwestern Polytechnical University, China. Each speaker has delivered the speech more than 30 minutes. In addition, around 50 experts and scholars and listeners from the world's top research institutes of aerospace technology, including Northwestern Polytechnical University, Nanjing University of Aeronautics and Astronautics, Beihang University, Nanjing University of Science and Technology.

We would especially like to thank the organization committees, the members of the program committees and technical committees. They have worked very hard in reviewing papers and making valuable suggestions for the authors to improve their work. We also would like to express our gratitude to the external reviewers, for providing extra helps in the review process, and the authors for contributing their research result to the conference. Without such support a conference cannot maintain quality or add real original contributions to knowledge. It goes without saying, it is core to success, but here we re-iterate that point.

On behalf of organizing committees, we would like to express our sincere gratitude to all participants and speakers who contribute to make this event a great success.

Conference Organizing Committee
AAAC 2023

Journal of Physics: Conference Series **2633** (2023) 011001 doi:10.1088/1742-6596/2633/1/011001

Organizing Committees

Conference Chairs
Renfu Li, Huazhong University of Science and Technology, China
Zheng Hong Zhu, York University, Canada

Program Chairs
Chih-Yung Wen, The Hong Kong Polytechnic University, Hong Kong, China
Xuelin Lei, East China University of Science and Technology, China

Program Committee
Yan Wang, Nanjing University of Aeronautics and Astronautics, China
Yunfei Zhang, Beihang University, China
Xingjian Jing, City University of Hong Kong, Hong Kong, China

Publicity Chairs
Rui Zhao, Beijing Institute of Technology, China
Danhe Chen, Nanjing University of Science and Technology, China

Technical Committee Members
Mohd Na'Im Abdullah, Universiti Putra Malaysia, Malaysia
Mahmut AdilCYÜKSELEN, Istanbul Technical University, Turkey
Luling An, Nanjing University of Aeronautics and Astronautics, China
Fathinul Fikri AS, Universiti Putra Malaysia, Malaysia
Dharmahinder Singh Chand, Chandigarh University, India
Qu Feng, Northwestern Polytechnical University, China
Chingiz Hajiyev, Istanbul Technical University, Turkey
Jiaao Hao, The Hong Kong Polytechnic University, Hong Kong, China
Fatih Karpat, Uludag University, Turkey
Metin Orhan Kaya, Istanbul Technical University, Turkey
Vsevolod V. Koryanov, Bauman Moscow State Technical University, Russia
Assoc. Prof. Ozlem Sahin, Eskisehir Technical University, Turkey
Ferhan Kuyucak Şengür, Anadolu University, Turkey
Junwei Wang, China Academy of Space Technology (CAST) , China
Zebbiche Toufik, Blida University 1, Algeria
Pavlo Maruschak, Ternopil Ivan Puluj National Technical University, Ukraine
Hu Ruiqin, Beijing Institute of Spacecraft Environment Engineering, China
Tahir Hikmet Karakoç, Anadolu University, Turkey
Hui Gao, Beihang University, China
Shengzhou Bai, The Hong Kong Polytechnic University, Hong Kong, China
Yuanyuan Zhang, Huazhong University of Science and Technology, China
Fengquan Zhong, Institute of Mechanics, Chinese Academy of Sciences, China
Linyuan Jia, Northwestern Polytechnical University, China
Fengnian Tian, Huazhong University of Science and Technology, China
Rui Zhao, Beijing Institute of Technology, China

Journal of Physics: Conference Series **2633** (2023) 011001 doi:10.1088/1742-6596/2633/1/011001

STATEMENT OF PEER REVIEW

All papers published in this volume of *Journal of Physics: Conference Series* have been peer reviewed through processes administered by the proceedings Editors. Reviews were conducted by expert referees to the professional and scientific standards expected of a proceedings journal published by IOP Publishing.

Peer Review Statement

All papers published in this volume have been reviewed through processes administered by the Editors. Reviews were conducted by expert referees to the professional and scientific standards expected of a proceedings journal published by IOP Publishing.

- **Type of peer review:** Double Anonymous
- **Conference submission management system:** Morressier, Zmeeting
- **Number of submissions received:** 56
- **Number of submissions sent for review:** 56
- **Number of submissions accepted:** 29
- **Acceptance Rate (Submissions Accepted / Submissions Received × 100):** 51.8
- **Average number of reviews per paper:** 1
- **Total number of reviewers involved:** 35
- **Contact person for queries:**
  Name: Xuelin Lei
  Email: xuelinlei@ecust.edu.cn
  Affiliation: East China University of Science and Technology, China

Research of a novel aerodynamic evaluation method for fixed-wing UAV

Yansen Wu[1], Deshan Liu[2], Tiancheng Wang[2] and Anmin Zhao[1,2] *

[1] School of Aeronautical Science and Engineering, Beihang University, 37 Xueyuan Road, Haidian District, Beijing, P.R, China
[2] Wing Flying (Tianjin) Technologies Co., Ltd, 7th Floor, B2#,
Animation Building, Sino-Singapore Tianjin ECO-CITY, China

* Corresponding author's email: zhaoanmin@buaa.edu.cn

Abstract. The aerodynamic parameters of unmanned aerial vehicles (UAVs) play an important role in the newly designed drone. For the purpose of the real-time measurement of the aerodynamic performance parameters of the UAV in flight status, based on the self-development connected-wing configuration airplane, a novel airborne aerodynamics test scheme is researched and verified in this work. The findings of this study indicate that the effectiveness of numerical calculations can be verified by comparison with both results of the numerical simulation and airborne flight test in the same conditions. The newly designed UAV equipped with this test system in the future, not only achieves quick and reliable results of aerodynamic performance and powered system efficiency but also can save the expense of wind tunnel experiments, which is of high potential to the UAV application.

1. Introduction

Unmanned aerial vehicles (UAVs) industry is booming along with its large quantity of engineering applications like hazardous environment monitoring, search and rescue, border surveillance and military community, etc [1]. It can be divided into three main streams of UAVs which are fixed-wing type, multirotor type, and hybrid type. In terms of the mission specifications and requirements, the three kinds of aerial vehicles are chosen to fit in different application domains depending on the requirement of performance capabilities including flight speed, flight endurance, payload capacity, etc [2].

To keep up with the targets for enhanced aerodynamic maneuverability and performance, novel technologies constantly emerge in the last decades such as novel configuration layouts, flow control techniques, and propulsion architectures, where predicting and obtaining aerodynamic parameters plays a significant role, especially in the investigation of those technologies and aircraft design. Therefore, it is urgently required for rapid and accurate methods to estimate or measure aerodynamic and stability coefficients and facilitate performance calculations.

Three conventional methods are capable of providing aerodynamic parameters [3, 4]. Firstly, the Theoretical calculation method that uses existing analytical theories and equations can provide reliable results for traditional configurations based on ideal hypotheses and proper simplifications during the conceptual design. Secondly, to obtain the final concept constraints, numerical methods are adopted in aerodynamic design like Computational Fluid Dynamics (CFD) simulations to achieve an accurate estimation of the aerodynamic coefficients. Finally, wind tunnel testing is an approach to measure the aerodynamic data in a simulated flying environment [5]. Hence, the emphasis of these methods is given to the accuracy and consistency with a real flight of the UAV aerodynamic figures.

In the present study, a large quantity of aerodynamic analysis methods has been presented and discussed. Many of them develop diverse models to calculate specific flight features of aircraft [6-8]. A transition prediction method based on the coupled Michel transition Criterion and $\gamma - Re\ \theta$ transition model has been proposed by Chen et al. for aerodynamic analysis of low-speed fixed-wing UAVs [9]. Shen et al. developed the procedure of computing longitudinal and lateral aerodynamic derivatives of Small-scaled Fixed-wing UAVs, aiming to improve the accuracy of the Aircraft Dynamic Model [10]. A simple but more comprehensive aerodynamic force model was developed by Bannwarth et al. to make a balance between high accuracy and computational Complexity [11]. Moreover, numerical methods are also used to estimate, identify or optimize aerodynamic parameters [12-17]. Piedra et al. verified the flight requirements of light sport aircraft numerically with the Vortex Lattice Method and Computational Fluid Dynamics simulations concluded[18]. Saderla et al. carried out an online aerodynamic parameter estimation method of UAV, which provide high-accuracy results compared with other methods [2, 19]. Analytical methods provide reliable results as well as convenient ways for estimating aerodynamic parameters. However, there still exists a non-negligible margin of error when compared with testing data obtained from wind tunnels and flight tests in the real flight environment. Thus, experimental methods are necessarily required to obtain and analyze the aerodynamic performance of UAVs to increase data accuracy.

It is previously found in the review that a large diversity of research on the aerodynamic parameter test has been studied currently. Nevertheless, the relevant investigations associated with the airborne test approach have hardly been reported. In this study, a novel airborne aerodynamic estimation system of unmanned aerial vehicles is constructed, which enables aircraft to achieve measurement of aerodynamic data in real flight environment conditions.

2. Aerodynamic parameter test system

Based on the developed electric propulsion test system, a UAV airborne aerodynamic parameter test system (AAPTS) is proposed and applied. It can not only obtain real-time power system efficiency parameters but also acquire real-time aerodynamic parameters of the aircraft in flight state.

2.1. A.Measurement theory equation

A measurement scheme has been implemented for the airborne aerodynamic system based on the 2.4GHz network transmission and real-time data processing. The parameters of the aerodynamic performance and the system efficiency of the aircraft, i.e., lift coefficient (abbr. CL), drag coefficient (abbr. CD), lift-drag ratio (abbr. i) can be derived based on the data provided by the measurement system, and Eq.(1)-(2).

Table 1. Measurable parameters of the airborne aerodynamic and electric power evaluation system

Item	Voltage	Current	Thrust	Torque	RPM	Airspeed
Symbol	U	I	F	T	n	V
Unit	V	A	gf	N · m	r/min	m/s

The aircraft is set to cruise in a steady state by utilizing a ground control station platform while the measurements are carried out. This assumption allows the establishment of a balance of forces in the direction of flight and the vertical direction, respectively.

$$\begin{aligned} \text{Lift(L)} &= \text{Weight(W)} \\ \text{Drag(D)} &= \text{Thrust(T)} \end{aligned} \tag{1}$$

Where the weight W is evaluated as a constant value owing to there being no fuel reduction for purely electric-powered aircraft. The thrust T of the propeller is obtained by the aerodynamic parameter test system.

The lift coefficient CL and drag coefficient CD are equal to the lift and drag force divided by the product of the corresponding airflow pressure and reference wing area respectively. The lift-to-drag ratio is calculated by dividing the CL by the CD, which is the same as dividing the lift equation by the drag equation.

$$C_L = \frac{L}{1/2\rho v^2 S} = \frac{W}{1/2\rho v^2 S}$$
$$C_D = \frac{D}{1/2\rho v^2 S} = \frac{T}{1/2\rho v^2 S} \qquad (2)$$
$$i = \frac{L}{D} = \frac{\text{Lift}}{\text{Drag}} = \frac{C_L}{C_D}$$

Where ρ is air density, v is the airspeed, and S the area of the wing. The ratio is perhaps the most significant consideration in designing a new UAV. And the higher the percentage, the better the aircraft will own aerodynamics.

2.2. Tested platform

To verify whether the aerodynamic parameters of the aircraft in flight state for the new proposed airborne aerodynamic parameters test system can be obtained, a self-development connected-wing airplane platform is employed. It has a mass of 12kg, a wingspan of 2.6m, and an area of 1.18 square meters. Detailed data that the aircraft's technical data & specifications are presented in Table 2. Because of the layout of the connected wing, the wing is divided into front and rear wings with the aspect ratio of 8.55 and 8.25 respectively.

Table 2. Physical specifications of the self-development connected-wing airplane

Wingspan	2600 mm	Weight	12 kg
Cruising speed	16.67 m/s	Body Length	1300 mm
Aspect Ratio of the front wing	8.55	Aspect Ratio of the rear wing	8.25
Wing Area	1.18 m2	Mean Aerodynamic Chord	304 mm

3. Verification and analysis

To verify the validity of the numerical calculation results, the airborne performance test of the UAV is carried out. Figure 1 shows the physical image of the test prototype (the third-generation composite material prototype). At the same time, in order to test the thrust and aerodynamic parameters of the connected wing aircraft in real-time flight conditions, an airborne dynamic tension test system was installed at the concentric location of the fixed-wing motors. The assembly is completed as shown in Figure 2.

Figure 1. The physical picture of the whole vertical take-off and landing UAV

Figure 2. UAV airborne flight performance test equipment

In order to complete this experiment, the experimental design elements are as follows: 1) Before the flight test, the airborne power test system has been checked and the flight control system has been debugged; 2) The flight path is a fixed-point hovering with a radius of 300 m; 3) In the fixed-height linear cruise state of the aircraft, the thrust is collected by the tension sensor, and the lift of the aircraft is equal to the gravity of the whole aircraft. The other electrical signals and power system parameters: voltage, current, airspeed, speed, torque, and other parameters are also recorded by the data acquisition system; 4) The angle of attack of the aircraft is recorded by the flight control system of the UAV.

Table 3 shows the dynamic pull test conditions of the connected-wing UAV. For the fully assembled UAV, the take-off weight is 12 kg. According to the different angles of attack of the UAV, 4 sets of flight tests are performed, and each set of tests is repeated 5 times. The test site was selected in the yurt airspace of Binhai New Area, Tianjin. Figure 3 is an image of the flight test process. Figure 4 shows the aerodynamic curves of the UAV within the range of the angle of attack. The comparison results show that the lift and drag coefficients of the airborne experiment and the numerical calculation are in good agreement. Therefore, it can also be concluded that the airborne performance test experiment also possible replace the traditional wind tunnel experiment to complete the verification of the numerical solution.

Figure 3. Snapshots of the aircraft during the test.

Table 3. Experimental conditions for airborne performance test of connected- wing UAV

Takeoff weight	Test times at a single angle of attack	Angle of attack
12 kg	5	0°, 2°, 4°, 6°

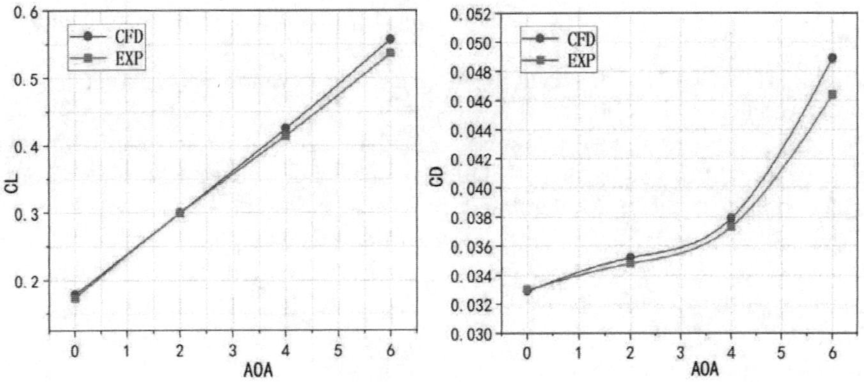

Figure 4. The aerodynamic performance of the connected wing aircraft at various AOA

4. Conclusions

The airborne dynamic tension test system proposed in this paper can conduct real-time online tests on the power system of electric UAVs and finally forms a complete test platform. Through the comparative analysis of numerical calculation and airborne test results, the following conclusions are obtained:

A novel aerodynamic evaluation method system of the fixed-wing is put forward in this work, and flight evaluations are performed to investigate the reliability of the newly proposed airborne efficiency system. Airborne experiments were conducted to verify the numerical calculation, which is potentially promising for electric aircraft optimization.

Acknowledgments

The work is supported by the China Postdoctoral Science Foundation (No. 2022M720349) and Wing Flying (Tianjin) Technologies Co., Ltd, Tianjin. for the financial and technical support (No. etcore-01-2020).

References

[1] AUSTIN R. Unmanned aircraft systems: UAVs design, development, and deployment [J]. 2010, 79(50): 31-6.

[2] SAEED A S, YOUNES A B, CAI C X, et al. A survey of hybrid Unmanned Aerial Vehicles [J]. Prog Aeosp Sci, 2018, 98(91-105.

[3] PEREIRA J D. Wind tunnels: aerodynamics, models and experiments [M]. Nova Science Publishers, 2011.

[4] LERNER J C, BOLDES U. Wind tunnels and experimental fluid dynamics research [M]. BoD - Books on Demand, 2011.

[5] YONDO R, ANDRES E, VALERO E. A review on design of experiments and surrogate models in aircraft real-time and many-query aerodynamic analyses [J]. Prog Aeosp Sci, 2018, 96(23-61.

[6] TUGNOLI M, MONTAGNANI D, SYAL M, et al. Mid-fidelity approach to aerodynamic simulations of unconventional VTOL aircraft configurations [J]. Aerosp Sci Technol, 2021, 115(17.

[7] CZYZ Z, KARPINSKI P. AERODYNAMIC CHARACTERISTICS OF THE X-TAIL STABILIZER IN A HYBRID UNMANNED AIRCRAFT [J]. Int J Simul Model, 2020, 19(4): 631-42.

[8] MA H. Calculation and Analysis of Aerodynamic Parameters of Wing Based on NS Equation [J]. Journal of Physics: Conference Series, 2019, 1300(012054.

[9] CHEN L L, GUO Z, DENG X L, et al. Aerodynamic performance and transition prediction of low-speed fixed-wing unmanned aerial vehicles in full configuration based on improved gamma - Re-theta model [J]. Aerosp Sci Technol, 2020, 107(15.

[10] SHEN J, SU Y, LIANG Q, et al. Calculation and Identification of the Aerodynamic Parameters for Small-Scaled Fixed-Wing UAVs [J]. Sensors, 2018, 18(2): 206.
[11] BANNWARTH J, CHEN Z J, STOL K A, et al. Aerodynamic Force Modeling of Multirotor Unmanned Aerial Vehicles [J]. 2019, 57(3): 1250-9.
[12] OLEJNICZAK D, NOWACKI M. Assessment of the selected parameters of aerodynamics for Airbus A380 aircraft on the basis of CFD tests; proceedings of the 13th International Scientific Conference on Sustainable, Modern and Safe Transport (TRANSCOM), Novy Smokovec, SLOVAKIA, F May 29-31, 2019 [C]. Elsevier: AMSTERDAM, 2019.
[13] ZHANG W, ZHU J. Online identification of aerodynamics with fast time - varying features using Kalman filter [J]. IET Control Theory & Applications, 2021, 15(2): 272-80.
[14] ABUDARAG S, YAGOUB R, ELFATIH H, et al. Computational analysis of unmanned aerial vehicle (UAV), F 2017]. Author(s).
[15] BUCKLEY H P, ZINGG D W J A J. Approach to Aerodynamic Design Through Numerical Optimization [J]. 2013, 51(8): 1972-81.
[16] BOUTEMEDJET A, SAMARDZIC M, REBHI L, et al. UAV aerodynamic design involving genetic algorithm and artificial neural network for wing preliminary computation [J]. Aerosp Sci Technol, 2019, 84(464-83.
[17] PANAGIOTOU P, DIMOPOULOS T, DIMITRIOU S, et al. Quasi-3D Aerodynamic Analysis Method for Blended-Wing-Body UAV Configurations [J]. Aerospace, 2021, 8(1): 13.
[18] PIEDRA S, MARTINEZ E, ESCALANTE-VELAZQUEZ C A, et al. Computational aerodynamics analysis of a light sport aircraft: Compliance study for stall speed and longitudinal stability certification requirements [J]. Aerosp Sci Technol, 2018, 82-83(234-42.
[19] SADERLA S, RAJARAM D, GHOSH A K. Parameter Estimation of Unmanned Flight Vehicle Using Wind Tunnel Testing and Real Flight Data [J]. Journal of Aerospace Engineering, 2017, 30(1): 04016078.

Performance optimization design of wide-body aircraft pressure refueling system

Hao Gong

Shanghai Aircraft Design and Research Institute, Commercial Aircraft Corporation of China, Ltd., Shanghai, 5188 Jinke Road, 201210, China

haogong2017@163.com

Abstract. In order to reduce the wide body aircraft refueling process time-cost and improve the airport operation efficiency, pressure refueling system design strategy was innovated. The shutoff valves' operation sequence in refueling process was optimized. The diameter of the orifices on the pressure refueling system fuel line was designed. Pressure refueling performance was simulated and analyzed at different scenarios. The results show that it can improve the pressure refueling system performance at different scenarios by optimizing the pressure refueling process strategy design including optimizing the fuel line orifice design and optimizing the wing tank inner side shut-off valve operation sequence.

1. Introduction

The pressure refueling process is an important part of the airliner ground service. In general, the refueling process occupies most of the ground operation time, especially for the long-range wide body aircraft. Shortening the refueling process time-cost is beneficial for airlines, airports, passengers, and others by improving civil aircraft operational efficiency and reducing airport working load [1-2].

Simultaneous refueling all the fuel tanks from empty or a certain value to full at the same time is usually chosen as the refueling system design condition. Besides. in order to meet the fuel line flow velocity limit [3], the fixed geometric orifices are designed and set on the refueling outlet line to limit the fuel flow rate, and the pressure refueling process is adjusted by the fuel gauging and controlling system. For the off-design condition, the refueling system performance will deviate from that of the design point. For example, when refueling only a portion of the tanks, i.e. not refueling all the tanks to the maximum quantity, it is not possible to refuel at the maximum refueling flow rate due to the fixed geometric orifices at the fuel line, and the time-cost of the refueling process can not be significantly reduced at this scenario. Actually, in most operation cases, the airliners refuel the aircraft fuel tanks according to the need of the flight task rather than refueling all the tanks to the maximum fuel quantity for economic consideration.

In order to increase the refueling flow rate, current technologies adopted in the wide-body aircraft are as follows.

(a) Using the refueling adapter with two nozzles, or setting two independent refueling adapters on one side of the aircraft wing and connecting them in parallel to the refueling line. It can increase the total refueling upstream flow rate of the refueling system.

(b) Installing the refueling adapters on both aircraft wings and adding some refueling pipes. It can increase the total refueling flow rate of both upstream and downstream of the refueling system.

As above technologies will increase the refueling system weight or cause some difficulties on the refueling system layout, design optimization of the refueling system is needed.

Many scholars have conducted relevant research on the aircraft pressure refueling system design optimization. Cao K Q optimized the aircraft ground pressure refueling system by installing variable geometric throttling devices in each refueling branch line to ensure the refueling time can be shortest for any refueling quantity [4]. Cao K Q conducted flow balance design for aircraft pressure refueling systems aimed at different actual refueling needs and control strategies [5]. Shen Y L used the Bayes method to optimize the design of the orifice geometry, which improved the efficiency of the aircraft pressure refueling system [6]. Wan F Q applied an improved BP neural network algorithm to fault-tolerant control of aircraft pressure refueling systems, which shortened refueling time and improved refueling control accuracy [7]. Wang J conducted simulation analysis on the aircraft ground pressure refueling system and calculated the refueling time and maximum flow rate in fuel line [8]. Pinheiro N conducted modeling and simulation of the fuel system based on AMESim [9]. Tu Y conducted modeling and simulation of the A320 aircraft refueling system based on Flowmaster, and analyzed the pressure distribution in the fuel line network [10]. Liu M developed a pressure refueling model based on Flowmaster, checked the impact pressure and ventilation capacity of the system [11]. Gong H proposed an innovative pressure refueling system from the perspective of control strategy [12].

In addition, due to the differences in policies and requirements of different countries, the aircraft refueling adaptor may not be set on the same side with the airport stand refueling station. If the refueling adaptor is quipped in only one wing, it may cause some inconvenience to the airline operation in the different countries.

This article optimized the working sequence of the refueling shut-off valves and the geometry of the refueling line orifices for a three-tank layout wide-body aircraft application. Simulation and analysis have been conducted on the different refueling scenarios including using a single side refueling adapter and using both sides of the refueling adapters simultaneously.

2. System description

This article analyzed the pressure refueling system for a three-tank layout wide body aircraft application. The system scheme is shown in Figure 1[12]. The mainly features of system are as follows.

(a) To adapt to the layout of ground refueling devices at different airports, one refueling station is set on the aircraft left wing and the other one is set on the right wing. Each refueling station is equipped with two standard refueling connectors, i.e. 2.5-inch couplings conforming to ISO45. The onboard refueling line is connected with the external refueling hoses (not shown in Figure 1) through the pressure refueling couplings.

(b) The aircraft's fuel storage tanks are divided into three independent parts including left-wing tank, central tank and right-wing tank. In order to ensure the reliability of the pressure refueling system, two pressure refueling outlets are set on the pressure refueling line for each independent fuel tank, and a shut-off valve and a orifice are installed on the refueling outlet line.

(c) Independent control shown in Table 1 is carried out on the refueling shut-off valves which are labeled as 3A, 3B, 3C, 3D, 3E, 3F in Figure 1. It can maintain a high refueling flow rate in various situations to shorten refueling time and ensure that the refueling flow rate does not exceed the limit value. When the fuel quantity of a certain fuel tank reaches the target value, the refueling shut-off valves inside the tank will be closed. As traditional design, the refueling shut-off valves are opened synchronously, and the restriction of refueling flow rate is achieved through the orifices. The operational logic of refueling shut-off valves is shown in Table 1. The differences of the operational logic of refueling shut-off valves 3B and 3E between the proposed innovative solution and the traditional solution are indicated by * in Table 1, which reflects different design ideas.

(d) When setting each tank refueling quantity, the wing tanks shall be prior filled. When refueling to all fuel tanks, the wing tanks and central tank starts refueling simultaneously.

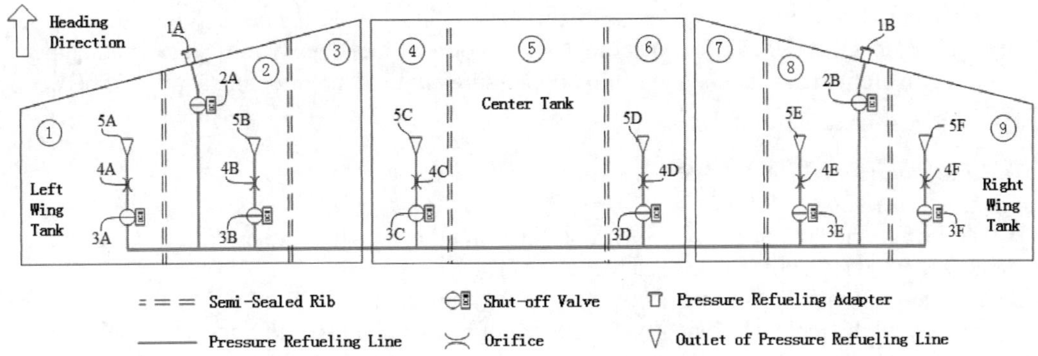

Figure 1. System technical scheme.

Table 1. Shut-off valve controlling scheme at different scenarios.

No.	Refueling adapters used	Fuel tank to be refueled			Shut-off valve condition					
		Left-wing tank	Center-wing tank	Right-wing tank	Valve 3A	Valve3B	Valve3C	Valve3D	Valve3E	Valve3F
1	One side	No	No	No	Closed	Closed	Closed	Closed	Closed	Closed
2	Both sides	No	No	No	Closed	Closed	Closed	Closed	Closed	Closed
3	One side	Yes	No	No	Opened	Opened	Closed	Closed	Closed	Closed
4	Both sides	Yes	No	No	Opened	Opened	Closed	Closed	Closed	Closed
5	One side	No	Yes	No	Closed	Closed	Opened	Opened	Closed	Closed
6	Both sides	No	Yes	No	Closed	Closed	Opened	Opened	Closed	Closed
7	One side	No	No	Yes	Closed	Closed	Closed	Closed	Opened	Opened
8	Both sides	No	No	Yes	Closed	Closed	Closed	Closed	Opened	Opened
9	One side	Yes	No	Yes	Opened	Opened	Closed	Closed	Opened	Opened
10	Both sides	Yes	No	Yes	Opened	Opened	Closed	Closed	Opened	Opened
11	One side	Yes	Yes	No	Opened	Closed*	Opened	Opened	Closed	Closed
12	Both sides	Yes	Yes	No	Opened	Opened	Opened	Opened	Closed	Closed
13	One side	No	Yes	Yes	Closed	Closed	Opened	Opened	Closed*	Opened
14	Both sides	No	Yes	Yes	Closed	Closed	Opened	Opened	Opened	Opened
15	One side	Yes	Yes	Yes	Opened	Closed*	Opened	Opened	Closed*	Opened
16	Both sides	Yes	Yes	Yes	Opened	Opened	Opened	Opened	Opened	Opened

Note, the content marked with * is only for proposed innovative design. For traditional design, it shall be "Opened".

3. System modeling

3.1. The design process

Based on the conceptual scheme shown in Figure 1 and Table 1, the design process of the pressure refueling system is shown in Figure 2. The different operational scenarios have been taken into account.

3.2. The design objectives

The design objectives are as follows.

(a) It takes no more than 40 minutes to refuel the central tank and the wing tanks from zero to maximum quantity at the pressure of 50 psig.

(b) It takes no more than 40 minutes to refuel the wing tanks from zero to maximum quantity at the pressure of 50 psig.

(c) It is better to shorten the time cost of the refueling process as much as possible.

3.3. The design constraints

The design constraints are as follows.

(a) The fuel flow velocity at any section of the refueling line shall not exceed 7 m/s.

(b) The fuel flow velocity at the outlet of the refueling line shall not exceed 1 m/s to avoid static electricity accumulation.

(c) At the end of refueling, the difference between the actual fuel quantity and the target refueling quantity shall be less than 100 kg for each fuel tank. A nominal fuel density of 0.8 kg/L is assumed.

(d) During the refueling process, the approved pressure range at the refueling adapter is between 35 psig to 55 psig.

3.4. The design scenarios

The design scenarios are as follows.

(a) To refuel all the fuel tanks using only one side refueling adapter, and prioritize ensuring that the wing tanks are filled to the maximum quantity.

(b) To refuel the wing tanks using a single side refueling adapter.

(c) To refuel all the fuel tanks using both sides of refueling adapters, and prioritize ensuring that the wing tanks are filled to the maximum quantity.

(d) The other scenarios in Table 1 are used to verify whether the design constraints are met.

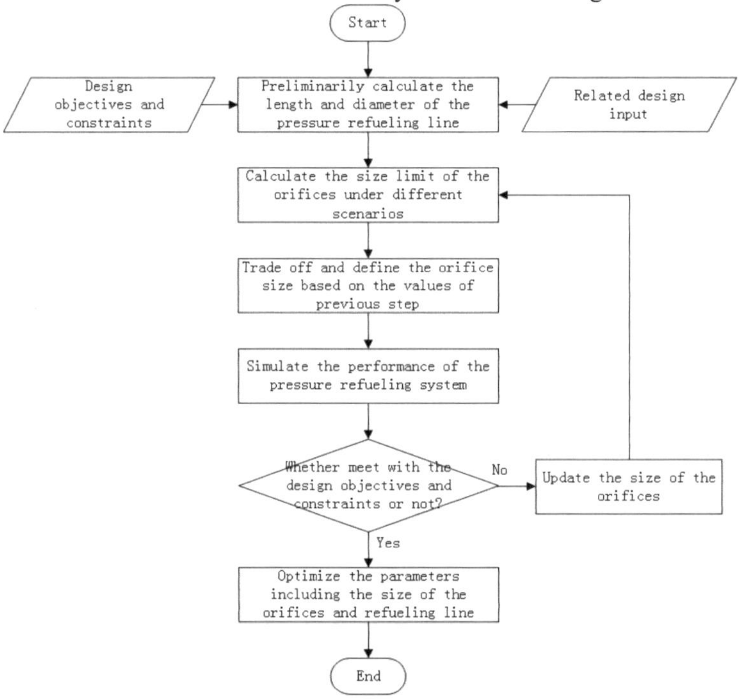

Figure 2. Pressure refueling system design process.

3.5. The performance model

The refueling line network architecture of the innovative refueling system design is the same with that of the traditional refueling system design. The differences between them lies in the design ideas, which lead to the differences of system design parameters such as the orifices' geometry and the control logic of the shut-off valves. As shown in Figure 3, a 1-D performance model of the pressure refueling line network was developed based on AMESim software. The model can be used to analyze the performance of the traditional refueling system design and the innovative refueling system design as shown in Figure 1 and Table 1.

Under the guidance of the process shown in Figure 2, the main parameters of the model have been calculated and defined. The results are shown in Table 2. The detail of the calculation process is not displayed here. Due to the fact that the maximum fuel capacity of the central tank is much greater than that of the wing tank, the refueling time of the central tank plays a decisive role in the total refueling process. The size of the orifices of central tank refueling line is similar for the traditional refueling system design and the innovative refueling system design. However, there are significant differences in the size of the orifices of wing tank refueling line between the traditional system design and the innovative system design.

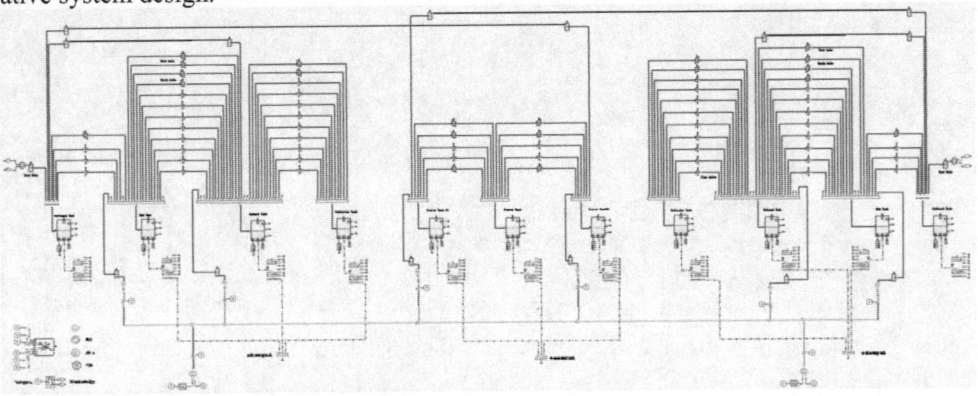

Figure 3. Pressure refueling 1-D fuel line performance model based on AMESim software.

Table 2. Pressure refueling system model parameter.

Category	Parameter	Unit	Value	
			Traditional design	Innovative design
left-wing tank	Maximum fuel quantity including the left collector tank	m^3	25.0	25.0
	Throat diameter of the orifice which is marked as 4A	mm	28.4	20.3
	Throat diameter of the orifice which is marked as 4B	mm	28.4	20.3
center-wing tank	Maximum fuel quantity	m^3	70.0	70.0
	Throat diameter of the orifice which is marked as 4C	mm	34.0	33.5
	Throat diameter of the orifice which is marked as 4D	mm	34.0	33.5
right-wing tank	Maximum fuel quantity including the right collector tank	m^3	25.0	25.0
	Throat diameter of the orifice which is marked as 4E	mm	28.4	20.3
	Throat diameter of the orifice which is marked as 4F	mm	28.4	20.3
Fuel line	Diameter of the fuel line in which refueling adapter is installed	mm	101.6	101.6
	Length of the fuel line in which refueling adapter is installed	m	3.0	3.0
	Diameter of main refueling line	mm	101.6	101.6
	Length of main refueling line	m	40.0	40.0
	Diameter of branch refueling line in central tank	mm	76.2	76.2
	Diameter of branch refueling line in wing tank	mm	50.8	50.8
	Length of branch refueling line	m	2.0	2.0

4. Simulation and discussion

Simulation on the pressure system performance at different refueling scenarios has been conducted based on the shut-off valve controlling scheme defined in Table 1. The simulation condition is to refuel Jet-A1 fuel to the target fuel tanks under 50 psig refueling pressure on a normal temperature day at sea level, and refueling process to all the target fuel tanks starts at the same time. Differences on pressure refueling performance between the innovative system design and the traditional system design are shown in Figure 4 to Figure 7.

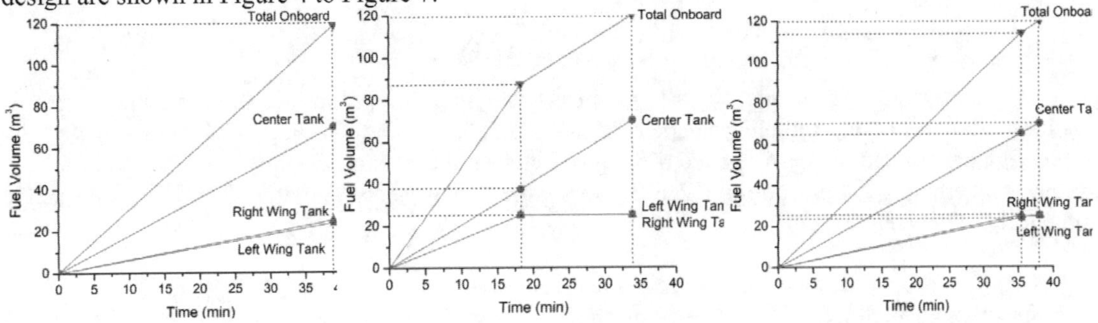

(a) Innovative design with one side adapter (b) Innovative design with both side adapters (c) Traditional design with one side adapter

Figure 4. Variation of onboard fuel quantity over time when refueling from empty to 120 m^3.

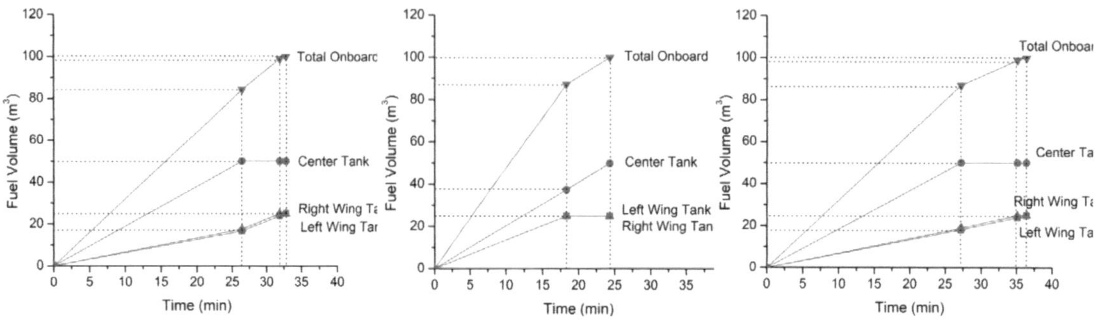

(a) Innovative design with one side adapter (b) Innovative design with both side adapters (c) Traditional design with one side adapter

Figure 5. Variation of onboard fuel quantity over time when refueling from empty to 100 m^3.

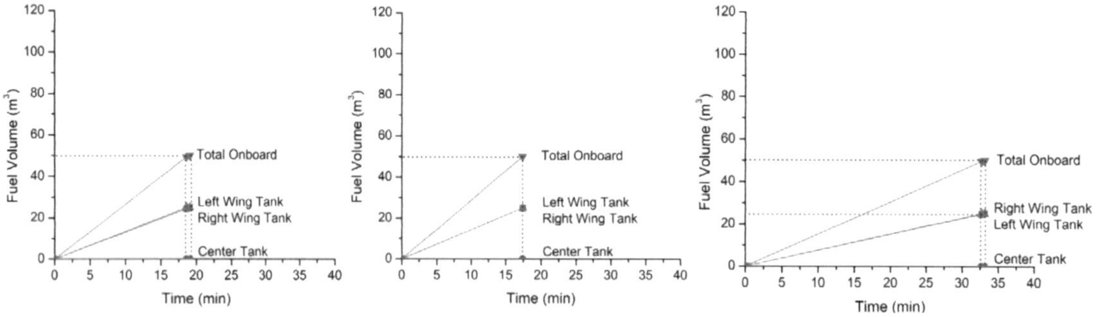

(a) Innovative design with one side adapter (b) Innovative design with both side adapters (c) Traditional design with one side adapter

Figure 6. Variation of onboard fuel quantity over time when refueling from empty to 50 m^3.

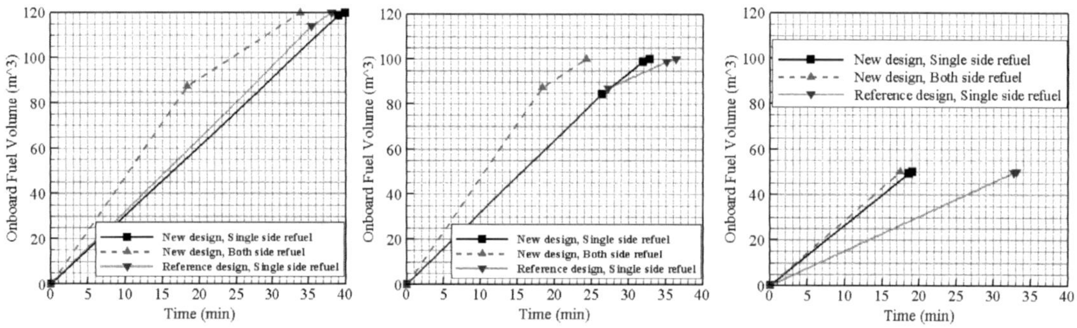

(a) Refueling to 120 m^3, i.e. all tanks full (b) Refueling to 100 m^3 (c) Refueling to 50 m^3, i.e. wing tanks full

Figure 7. Variation of onboard fuel quantity over time when refueling from empty.

Calculation results show that, for the traditional refueling system design, the refueling pattern of using single side of the refueling adapter is basically consistent with that using both sides of the refueling adapter, and the performance difference between them is relatively small. Therefore, for the traditional refueling system design, Figure 4 to Figure 7 only show the situation when using single side of refueling adapter.

4.1. Refueling process description for the innovative design

For the innovative design which has been shown in Figure 1 and Table 1, three commonly used automatic refueling scenarios were taken as examples to describe the refueling process.

(a) When refueling all the fuel tanks automatically by using the right side refueling adapter, and prioritize ensuring that the wing tanks are filled to the maximum quantity, the refueling process is shown in Figure 5 (a) and can be described as follows. Firstly, open the shut-off valves 3A, 3C, 3D and 3F on the refueling line, and keep the shut-off valves 3B and 3E closed. Secondly, start to refuel both wing tanks and central tank at the same time, and continue the refueling process until the central tank reaches the preset value. And then, close the shut-off valves 3C and 3D. Thirdly, continue to refuel the wing tanks until the fuel level of the wing tanks reaches the preset value. And then close the shut-off valves 3A, 3B, 3E and 3F.

(b) When refueling only the wing tanks automatically by using the right side refueling adapter, the refueling process is shown in Figure 6(a) and can be described as follows. Firstly, open the shut-off valves 3A, 3B, 3E and 3F on the refueling line, and keep the shut-off valves 3C and 3D closed. Secondly, start to refuel both wing tanks at the same time, and continue the refueling process until the wing tanks reach the preset value. And then, close the shut-off valves 3A, 3B, 3E and 3F.

(c) When refueling all the fuel tanks automatically by using the both side refueling adapters, and prioritize ensuring that the wing tanks are filled to the maximum quantity, the refueling process is shown in Figure 4(b) and Figure 5(b), and can be described as follows. Firstly, open the shut-off valves 3A, 3B, 3C, 3D, 3E and 3F on the refueling line. Secondly, start to refuel both wing tanks and central tank at the same time, and continue the refueling process until the wing tanks reach the preset value. And then, close the shut-off valves 3A, 3B, 3E and 3F. Thirdly, keep the shut-off valves 3C and 3D opened, and continue to refuel the central tank until the fuel level of the central tank reaches the preset value. And then close the shut-off valves 3C and 3D.

4.2. Refueling performance analysis

Based on the analysis to Figure 4 to Figure 7, it can be seen that the restrictions on refueling flow rate come from the following three aspects.

(a) The geometry of the orifice on the outlet of the branch refueling line. The throttling effect of the orifice can limit the fuel flow rate in the refueling line to a certain range, ensuring that the flow velocity in the refueling line does not exceed the limit under any possible operating condition. The optimization design for the orifices was carried out under the consideration that ensuring the flow velocity does not exceed the limit and increasing the total refueling flow rate as much as possible. In the scenario of maximum refueling quantity, the difference of the maximum refueling performance between the innovative design and the traditional design is relatively small, as shown in Figure 4 and Figure 7(a).

(b) The number of the refueling line outlets. On the premise that the size of the orifices remain unchanged and the number of refueling adapters has not yet reached the limit of further increasing the fuel flow rate, the fuel flow correspondingly increases when the number of refueling outlet increases. Taking Figure 6 and Figure 7(c) as an example, when only the wing tanks are refueled, it takes more than 30 minutes to complete the refueling process from zero fuel to maximum quantity for the traditional system. However, it can significantly shorten the refueling process time to within 20 minutes for the innovative system.

(c) Number of the refueling adapters. With the number of the refueling adapter increases, the refueling flow rate correspondingly increases, but the fuel flow rate is also constrained by the number of the refueling line outlets and the orifice' s geometry. If the number of refueling line outlet increases, the total refueling flow rate can be significantly increased for the innovative system design when using both wing sides of refueling adapters to refuel all the fuel tanks, as shown in Figure 5. However, for the innovative system design that only the wing tanks are refueled (as shown in Figure 6) and the traditional system design, there are limitations in improving the overall refueling flow rate if only the number of refueling adapter increases. That is due to the constant number of downstream refueling line outlets and the size of orifices on the refueling line. But it can reduce the fuel flow velocity in the refueling line section where the refueling adapters are located. In addition, setting the refueling adapters on both the left and right wings or setting a refueling adapter on one side as an aircraft optional item is beneficial for improving refueling operation flexibility.

5. Conclusion

This article conducts pressure refueling system design and analysis for a three-tank layout wide body aircraft. The main conclusions are as follows.

(a) The refueling flow rate can be maintained high in different refueling scenarios by optimizing the design of pressure refueling system, especially the control logic of the shut-off valves in the inner side of the wing tank. It can improve the operating performance of the refueling system.

(b) Optimizing the refueling strategy can reduce the refueling process time cost in the conditions that the central tank is not fully filled or only the wing tanks are refueled. It can improve the operational efficiency of airlines and airports.

(c) Setting the refueling adapters on both wings of aircraft or setting one side of refueling adapter as an aircraft optional item can improve the operational adaptability of the aircraft at different airports.

Acknowledgments

The author would like to acknowledge Zhu D X, Zhang D and Dong L K at Shanghai Aircraft Design and Research Institute for their suggestions to this article.

References

[1] Langton R, Clark C, Hewitt M and Richards L 2009 *Aircraft Fuel System* (Wiltshire: JohnWiley & Sons Ltd.)

[2] Tucci H N P, Neto G C O, Rodrigues F L, Giannetti B F and Almeida C M V B 2021 Six sigma with the blue economy fundamentals to assess the economic and environmental performance in the aircraft refueling process *Renewable and Sustainable Energy Reviews* **150** pp 1-14

[3] SAE International 2019 *Aircraft Fuel System Design Guidelines* AIR7975 (New York: SAE International)

[4] Cao K Q, Li X Y, Shen Y L, Wang J P and Hu L M 2004 Study on optimal design for big aircraft ground pressure refueling system *Machine Tool & Hydraulics* **2** pp 77-78

[5] Cao K Q, Liu B, Zhang C S and Yang X S 2006 Study of methods of flow matching design for aircraft ground pressure refueling system *J. Air Force Engineering University (Natural Science Edition)* **7** pp 4-6

[6] Shen Y L, Li Y L, Hu L M and Yang X S 2007 Optimal design for multitask complicated ground pressure refueling system based on Bayes method *J. System Simulation* **19** pp 905-8

[7] Wan F Q and Xiao N 2007 Reformative BP arithmetic in simulation and experiment of aircraft pressure refueling system *Aviation Precision Manufacturing Technology* **43** pp 57-62

[8] Wang J and Lu W 2014 Study on simulation of airplane pressure refueling system *Aeronautical Science & Technology* **25** pp 51-55

[9] Pinheiro N and Carlos L 2017 Modeling and simulation of a single engine aircraft fuel system *Proc. of the 15th Scandinavian Int. Conf. on Fluid Power* (Linköping: Linköping University) pp 45-50

[10] Tu Y, Chen X R, Yin L, Zheng Q Y and Zeng Y 2022 Numerical study on pipeline pressure surge of the large aircraft fuel system *Shock and Vibration* **2022** pp 1-13

[11] Liu M, Wang Z and Ren J 2022 Study on impact pressure and ventilation capability of aircraft pressure refueling system *CSAA/IET Int. Conf. on Aircraft Utility Systems* (Nanchang, China)

[12] Gong H, Zhang X Y, Dong L K and Zhang D 2023 Pressure refueling system and aircraft *CN Patent 218400999*

Wide-body aircraft fuel jettison system design and analysis

Hao Gong

Shanghai Aircraft Design and Research Institute, Commercial Aircraft Corporation of China, Ltd., Shanghai, 5188 Jinke Road, 201210, China

haogong2017@163.com

Abstract. In order to design the highly integrated fuel jettison system, the hydrodynamic performance design flow chart for the highly integrated fuel system has been proposed, and the 1-D fuel line network performance model has been developed. An innovative fuel jettison system for wide-body aircraft has been designed. The system performance has been simulated and analyzed. The results show that the proposed flow chart can be used for the highly integrated fuel system design, and the fuel jettison performance requirements can be met. The system parameters including the wing tank main pump characteristic, the flow area of the orifice set in the fuel line connecting the feeding line and the refueling line, and the flight height have significant effects on the fuel jettison performance. Besides, the aircraft total fuel jettison performance is also affected by the central tank override pump characteristic, the engine fuel consumption, and the flow resistance of fuel line network.

1. Introduction

The wide-body aircraft is widely used in the long range airlines, and its onboard fuel quantity is quite large. When some emergency cases happen, the aircraft is necessary to reroute and land off as soon as possible. To ensure the aircraft safety during the landing off process, it can reduce the aircraft weight within the limited time by jettisoning fuel in the airspace which is permitted by the government [1-2]. However, the frequency of jettisoning fuel is quite low in an aircraft full life cycle. To jettison fuel quickly and safely is the key problem to design the fuel jettison system. How to control the weight and cost increment caused by setting the fuel jettison system need to be paid attention and studied.

Many scholars have studied the aircraft fuel jettison system. Li M Z proposed a fuel jettison system design for transport aircraft, and analyzed the layout of the fuel line and jettison nozzle [3]. Chen Z B analyzed the requirements of the airworthiness article and studied the flight test method for the fuel jettison system of civil aircraft [4]. Wang P studied and summarized the legislative background, the airworthiness requirements, and recommended compliance method for civil aircraft jettison system [5]. Chen G H used a numerical method based on discrete phase model to simulate the wake pattern of fuel jettison, and studied the wake shape and motion trajectory under different pitch and yaw angles [6]. Yang A J discussed the gravity fuel jettison of wing tank, established the fuel jettison mathematical model based on the Flowmaster platform, and studied the key factors that affect the fuel jettison efficiency [7]. Luo J J designed an aircraft fuel jettison test system which can be used to test and verify the performance of the fuel jettison system [8]. Gong H raised a system concept design for correcting fuel imbalance between the aircraft wing tanks, the system can also support to jettison fuel [9].

This article proposed a hydrodynamic performance design flow chart for the highly integrated fuel system, and designed an innovative fuel jettison system for wide-body aircraft application based on the design flow chart. The 1-D fuel line network performance model was developed and the system performance was simulated and analyzed.

2. System description

Figure 1 showed a typical three tank layout wide-body aircraft engine feeding system and fuel jettison system. It reuses the components and pipelines of the engine feeding system and refueling system for fuel jettison. When jettisoning, all the wing tank main pumps and the central tank override pumps are involved in the fuel jettison process. To ensure the fuel supplied to the engine can meet the requirements, the orifices which are marked as 1 and 2 in Figure 1 are installed in the wing tank fuel line which connects the feeding line and the refueling line. The advantage of the design shown in Figure 1 is that the aircraft weight gained by the installation of the fuel jettison system can be limited. However, the engine feeding system and the fuel jettison system can not been isolated. In the event that the jettison shut-off valve cannot be closed due to a malfunction, there is a risk to shut off all the fuel pumps in one wing tank to stop jettison, which will affect the safety of engine feeding.

In response to the shortcomings of the system scheme shown in Figure 1, an innovative system architecture is proposed, which is shown in Figure 2. The two designs are based on the same engine feeding architecture and pressure refueling system, and the fuel jettison of the central tank is the same. The differences between them are that, the innovative design shown in Figure 2 has adjusted the connection position of the wing tank fuel jettison pipeline and the engine feeding pipeline, and a check valve is added at downstream of the main pump output pipeline. Due to the check valve, only one of the two main pumps in one wing tank can be used for jettison, while the other main pump is only used for engine feeding. Therefore, the isolation between engine feeding and fuel jettison can be achieved.

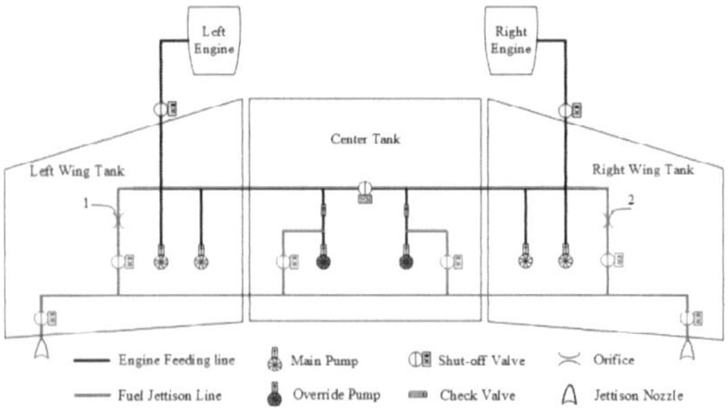

Figure 1. Typical engine feeding and fuel jettison system scheme.

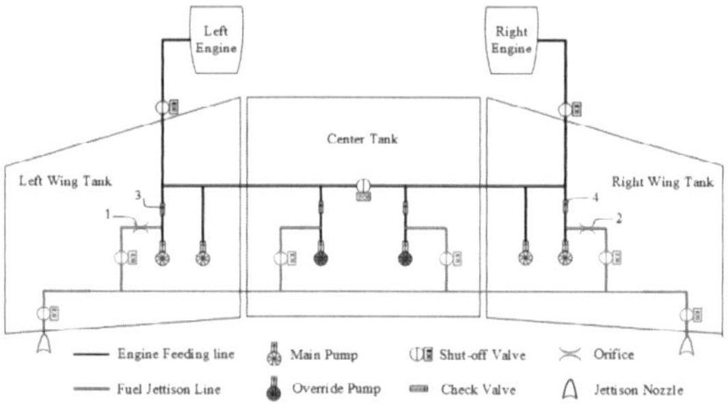

Figure 2. Innovative engine feeding and fuel jettison system scheme.

3. System modeling

For the highly integrated fuel system, there is a close cross-linking relationship between the hydrodynamic performance of different functions, as shown in Figure 3. This article proposed a design flow chart for the highly integrated fuel fluid system, as shown in Figure 4.

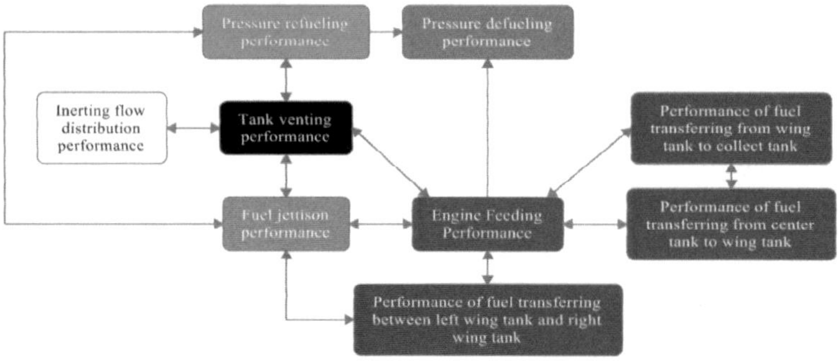

Figure 3. Relationship between the hydrodynamic performance of highly integrated fuel system.

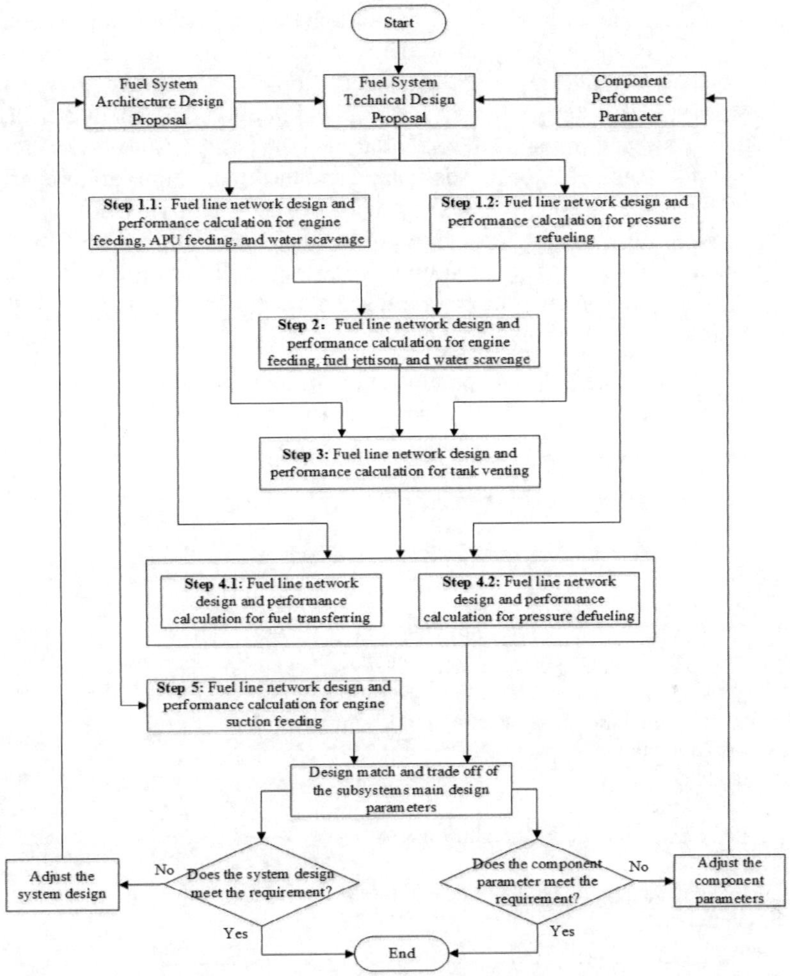

Figure 4. Flow chart of the hydrodynamic performance design for highly integrated fuel system.

In order to design and analyze the system performance, simulation models have been developed based on AMESim software for the typical architecture fuel system shown in Figure 1 and the innovative architecture fuel system shown in Figure 2. The two models are almost the same except for the connection position of the fuel jettison pipeline and the engine feeding pipeline in the wing tank, and a check valve is added at downstream of a main pump output pipeline for the innovative system. The AMESim model shown in Figure 5 is for the innovative system shown in Figure 2.

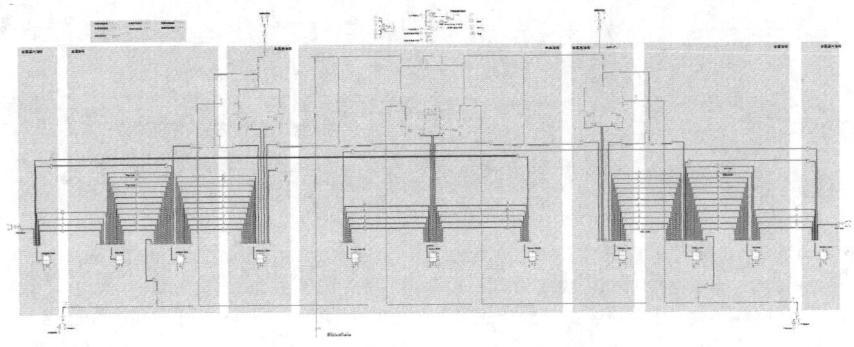

Figure 5. Engine feeding and fuel jettison performance model for the innovative design.

According to the process shown in Figure 4, the system design iteration was carried out to obtain the main parameter values of the hydrodynamic performance model for the fuel system. The design result is shown in Table 1. During the design and calculation process, the environmental parameters were based on the ISA standard day, and the fuel type was Jet A1. The characteristic curves of the main pump and the override pump assumed by similar equipment are respectively shown in Figure 6 and Figure 7. If the performance of the override pumps and the main pumps are properly matched, all the tanks can jettison fuel simultaneously. The flow ratio between the driving flow and the output flow (i.e. the sum of the driving flow and the input flow) of the injection pump was assumed to be 1:3. The output flow of the injection pump was set as similar to the engine fuel consumption. Considering that the research object of this article is only the engine feeding system and fuel jettison system, the design of other subsystems are not discussed in this article.

Table 1. Model parameter at design condition.

Parameter	Unit	Value	
		Typical design	Innovative design
Flight height [a]	m	10668	10668
Fuel temperature	℃	15	15
Engine fuel consumption	kg/s (per engine)	1.1	1.1
Aircraft weight reduction target [b]	kg	60000	60000
Maximum theoretical fuel jettison quantity of central tank	kg	40000	40000
Maximum theoretical fuel jettison quantity of single wing tank	kg (per wing)	10000	10000
Time-cost of jettison process [c]	min	≤60	≤60
Diameter of the orifice set in the fuel line connecting the feeding line and refueling line in the wing tank	mm	21.59	21.59
Engine feeding line diameter	mm	50.8	50.8
Main refueling line diameter	mm	88.9	88.9
Diameter of the fuel line connecting the feeding line and the refueling line in the wing tank	mm	50.8	50.8
Diameter of the fuel line connecting the feeding line and the refueling line in the central tank	mm	76.2	76.2
Fuel jettison line diameter	mm	50.8	50.8

[a] The flight height is chosen as the design condition, while the aircraft can jettison fuel at a range of flight height.
[b] The aircraft weight reduction is caused by both fuel jettison and engine fuel consumption.
[c] It is the time-cost for reducing aircraft weight.

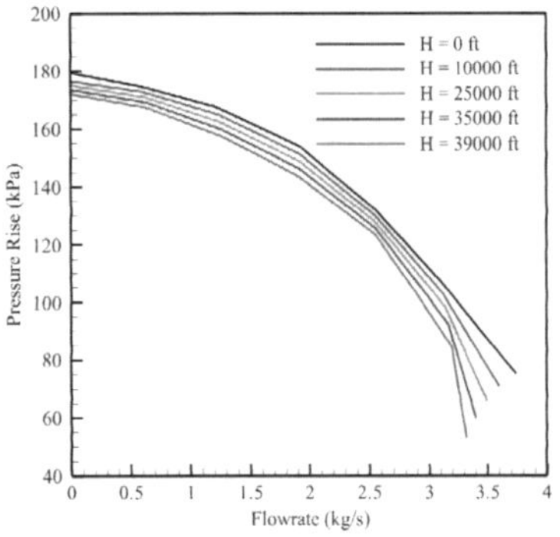

Figure 6. Characteristic curve of main pump.

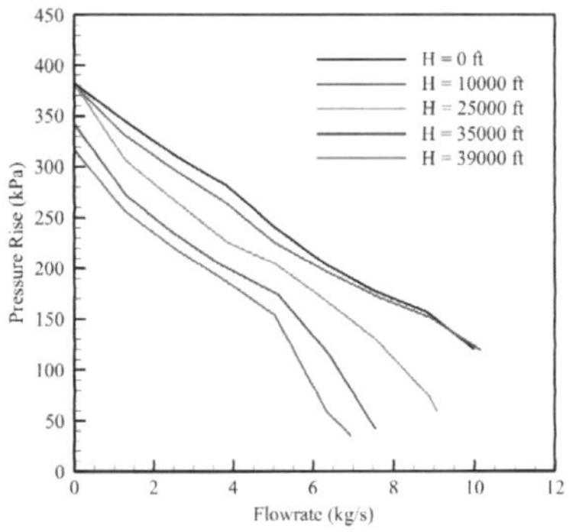

Figure 7. Characteristic curve of override pump.

The institute the author works for has built an aircraft engine feeding and fuel jettison test bench which can be referred to [8]. Based on the test bench and the parameters shown in Table 1, engine feeding and fuel jettison tests for the system architecture shown in Figure 1 have been conducted. The experimental data and the simulation results was compared in Table 2. According to engineering experience, the calculation accuracy of this simulation model can support the system design.

Table 2. Comparison the design condition simulation result with the test data for the typical design.

Parameter	Unit	Simulation result	Test data	Difference
The pressure at the interface of the feed line and the engine	kPa	122.1	116.5	4.80%
The engine fuel consumption	kg/s	1.11	1.10	0.91%
The pressure of the upstream side of the orifice set in the fuel line connecting the feeding line and the refueling line in the wing tank	kPa	116.1	110.8	4.78%
The fuel jettison flowrate of the single wing tank	kg/s	2.76	2.78	-0.72%
The pressure of the upstream side of the fuel jettison shut-off valve near the jettison line outlet	kPa	30.0	29.1	3.09%
The fuel flowrate at the jettison line outlet	kg/s	9.75	9.93	-2.01%

4. Simulation and discussion

4.1. Fuel jettison performance at the design condition

The fuel jettison performance of the typical architecture system and the innovative architecture system were simulated, the results are shown in Table 3. The parameter values used in the calculation referred to Table 1. From the data in Table 3, it can be seen that the performance of innovative architecture system is relatively close to that of typical architecture system. Taking fuel jettison at 10668 meter flight altitude as an example, the aircraft weight reduction target can be achieved by simultaneously jettisoning fuel from three fuel tanks in less than 50 minutes based on the data in Table 3.

Table 3. Fuel jettison performance of typical design and innovative design.

Condition	Parameter	Unit	All the tanks jettison at the same time		Only wing tanks jettison		Only central tank jettison	
			Typical design	Innovative design	Typical design	Innovative design	Typical design	Innovative design
Flight height: 10668m, Engine fuel consumption: 1.10kg/s, Number of operating engines: 2	Pressure at the interface with engine	kPa	122.10	144.75	105.10	144.91	167.60	165.93
	Left wing tank fuel jettison flowrate	kg/s	2.77	2.73	3.48	3.21	0.00	0.00
	central tank fuel jettison flowrate	kg/s	13.96	13.96	0.00	0.00	14.71	14.71
	Right wing tank fuel jettison flowrate	kg/s	2.77	2.73	3.48	3.21	0.00	0.00
	Total fuel jettison flowrate	kg/s	19.50	19.42	6.96	6.42	14.71	14.71
	Aircraft weight reduction rate	kg/s	21.70	21.62	9.16	8.62	16.91	16.91
Flight height: 4572 m, Engine fuel consumption: 1.78 kg/s, Number of operating engines: 2	Pressure at the interface with engine	kPa	130.08	132.80	100.12	131.67	171.56	169.32
	Left wing tank fuel jettison flowrate	kg/s	2.29	2.52	3.43	3.37	0.00	0.00
	central tank fuel jettison flowrate	kg/s	19.32	19.18	0.00	0.00	20.52	20.52
	Right wing tank fuel jettison flowrate	kg/s	2.29	2.52	3.43	3.37	0.00	0.00
	Total fuel jettison flowrate	kg/s	22.88	24.22	6.86	6.74	20.52	20.52
	Aircraft weight reduction rate	kg/s	26.44	27.78	10.42	10.30	24.08	24.08

4.2. Effects of the main pump characteristic and the wing tank jettison orifice flow area on the fuel jettison performance

The main pump pressure rise shown in Figure 6 is regarded as the reference value. In order to analyze the impact of the main pump characteristic on fuel jettison performance, five set of main pump pressure rise characteristic were formed by scaling based on the reference value. The scaling factor were 50%, 75%, 100%, 125%, and 150%. Simulation was conducted based on the main pump characteristic in conjunction with the override pump characteristic shown in Figure 7. It should be noted that, simulation analysis has shown that all five set of the main pump pressure rise can meet the engine feeding requirements, and ensure the central tank override feeding performance when there is no tank jettisoning.

The orifice discussed in this section refers to the flow limiting component which is installed on the fuel line connecting the engine feeding line and the refueling line in the wing tank, and it is used to limit the fuel jettison flow rate of the wing tank. The impact of the ratio of the flow area of the orifice to the cross-sectional area of the fuel line on fuel jettison performance was discussed in this section.

Figure 8 and Figure 9 show the fuel jettison performance of the innovative architecture system under different main pump pressure rise and wing tank jettison orifice sizes. The simulated flight altitude was 10668m Among them, Figure 8 shows the situation where all the tanks are simultaneously jettisoning, and Figure 9 shows the situation that only the wing tanks are jettisoning.

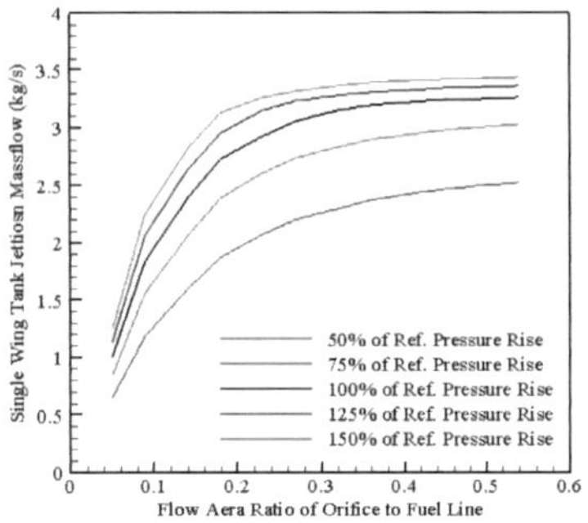

(a) Single wing tank jettison performance

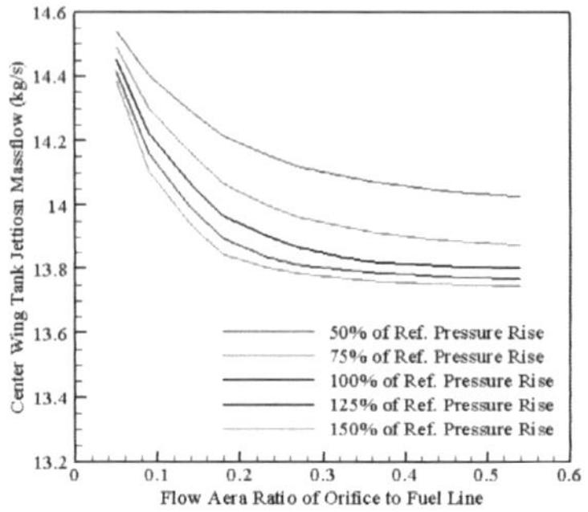

(b) central tank jettison performance

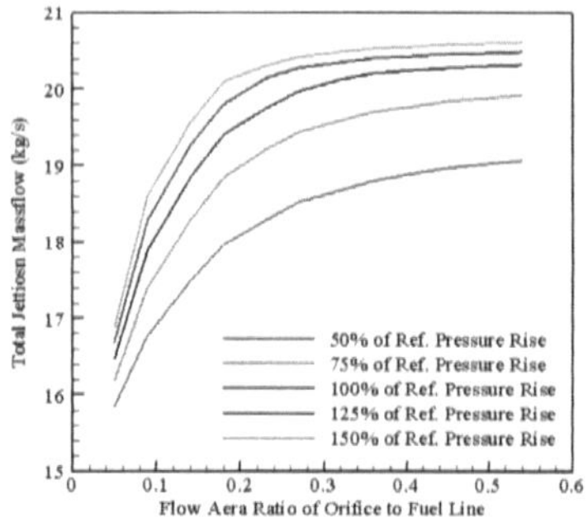

(c) Total jettison performance

Figure 8. Jettison performance with different main pump characteristic when all the tanks jettison.

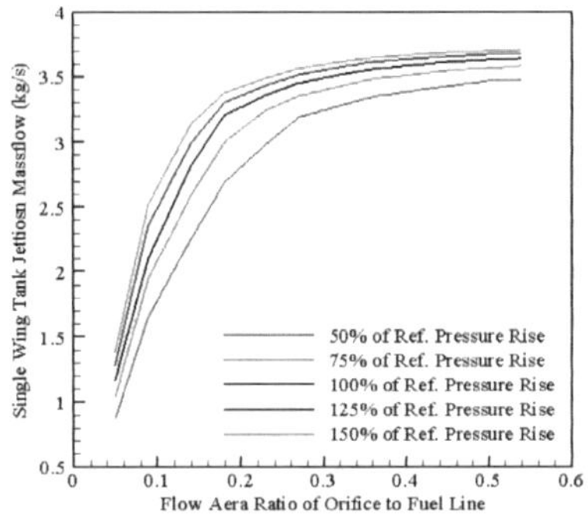

(a) Single wing tank jettison performance

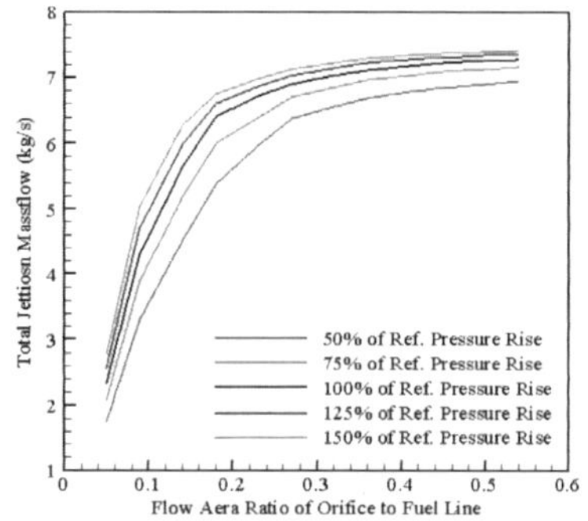

(b) Total fuel jettison performance

Figure 9. Jettison performance with different main pump characteristic when only wing tanks jettison.

As shown in Figure 8(a), when the ratio of the orifice flow area to the fuel line cross-sectional area increases, i.e. the flow area of the orifice increases, the wing tank fuel jettison flow rate increases correspondingly. When the ratio is small (such as between 0.05 and 0.30), the wing tank fuel jettison rate rapidly increases with the increase of the ratio. When the ratio is large, the wing tank fuel jettison flow rate increases slightly with the increase of this ratio. As reflected in Figure 8, the slope of the curve decreases with the increase of the x-axis parameter, that is, the larger the ratio, the smaller the corresponding increasing degree of the wing tank fuel jettison flow rate. In addition, as shown in Figure 8, under the same orifice flow area, the larger the main pump outlet pressure rise, the stronger the corresponding wing tank fuel jettison performance.

In the case of three fuel tanks jettison simultaneously, due to the convergence of the wing tank jettison flow and the central tank jettison flow at the downstream of the fuel jettison line, the fuel flow in the different fuel lines affects each other. Figure 8(a) and Figure 8(b) show that, as the flow area of the wing tank jettison connecting line orifice increases, the wing tank fuel jettison flow rate increases, but the central tank fuel jettison flow rate decreases. The relationship between the wing tank jettison flow rate and the central tank jettison flow rate is influenced by the flow resistance of the fuel jettison line. Besides, it is also related to the output performance of the main pump and the override pump. As shown in Figure 8(b), under the same orifice flow area, the larger the main pump pressure rise, the weaker the central tank fuel jettison performance. The total fuel jettison flow rate under the different orifice flow area is shown in Figure 8(c). Under the characteristics of the main pump and override pump in this article, the total fuel jettison flow rate increases with the increase of the orifice flow area, that is because the increase of the wing tank fuel jettison flow rate is stronger than the decrease of the central tank fuel jettison flow rate. Under the same orifice flow area, the larger the main pump pressure rise, the stronger the corresponding total fuel jettison performance.

Figure 9 shows the situation where only the wing tanks jettison fuel. At this time, the central tank override pumps do not work, and the change of the fuel jettison performance is only caused by the change of the wing tank fuel jettison performance. The parameter variation pattern in Figure 9(a) is the same as that in Figure 8(a). However, the fuel jettison flow rate in Figure 9(a) is slightly higher than that in Figure 8(a). That is because when only wing tanks jettison fuel, the jettison performance is affected by the wing tank fuel jettison line flow resistance, and is not affected by the central tank override pump performance. Compared to the three fuel tanks simultaneous fuel jettison, the downstream flow resistance is smaller at this time.

4.3. Effects of the flight height and the wing tank jettison orifice flow area on the fuel jettison performance

The purpose of the fuel jettison is to quickly reduce the weight of the aircraft, but it should be noted that the actual target is the reduction of the total weight of the aircraft, rather than the weight reduction caused only by fuel jettison. Figure 10 and Figure 11 show the innovative architecture jettison system performance at different flight altitudes with the reference fuel pump pressure rise characteristic. Figure 10 is for the situation of the three fuel tanks simultaneously jettison, and Figure 11 is for the situation of only the wing tanks jettison. The total aircraft weight reduction for the two situations is comprehensively shown in Figure 12.

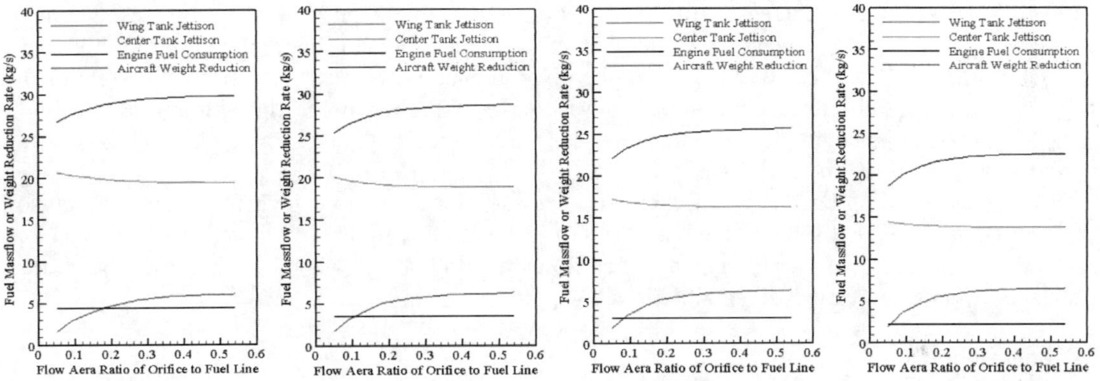

(a) 5000 ft (b) 15000 ft (c) 25000 ft (d) 35000 ft

Figure 10. The innovative architecture jettison system performance at different flight altitudes when all the tanks jettison simultaneously.

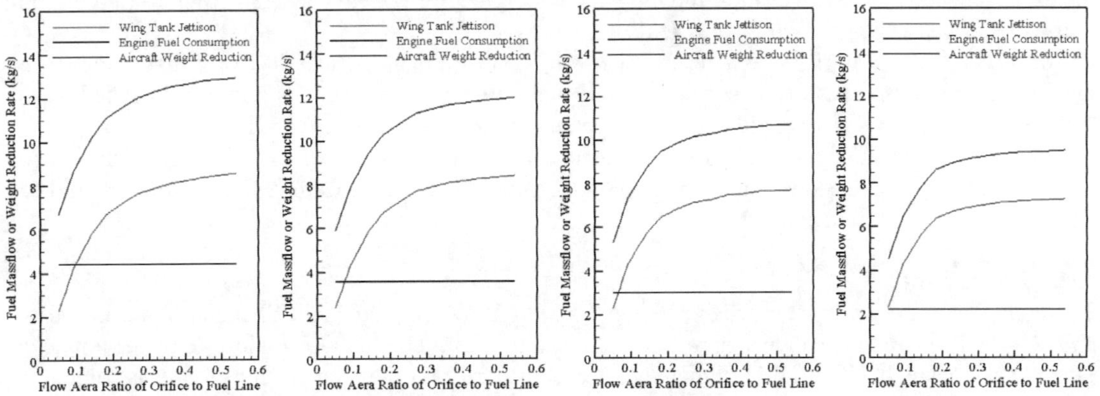

(a) 5000 ft (b) 15000 ft (c) 25000 ft (d) 35000 ft

Figure 11. The innovative architecture jettison system performance at different flight altitudes when only the wing tanks jettison.

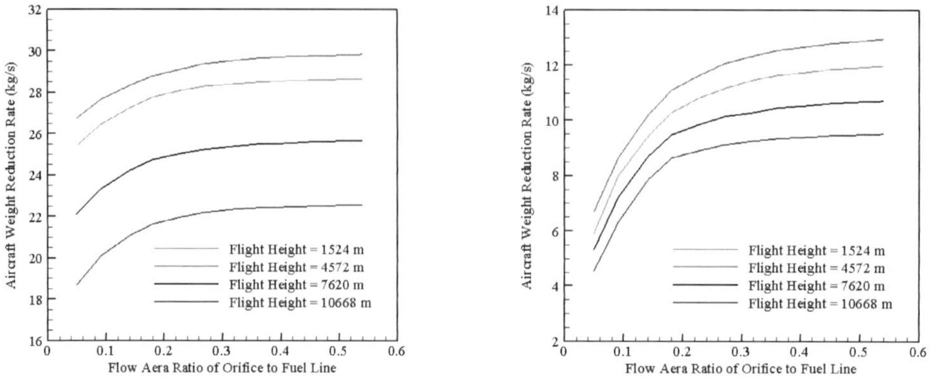

(a) All the tanks simultaneously jettison (b) Only the wing tanks jettison.

Figure 12. The total aircraft weight reduction at different flight altitudes.

As shown in Figure 10, at the given flight altitude, as the flow area of the wing tank jettison connecting line orifice increases, the wing tank jettison flow rate increases while the central tank jettison flow rate slightly decreases, finally the combined fuel jettison flow rate increases. The total aircraft weight reduction rate increases after adding the engine fuel consumption rate. From Figure 10 and Figure 12(a), it can be seen that, under the given orifice flow area, the jettison flow rate of the wing tank and central tank decreases with the increase of flight altitude. In addition, as the flight altitude increases, the engine fuel consumption decreases. Finally, the rate of weight reduction of the aircraft decreases as the flight altitude increases.

When only the wing tanks jettison, there is no cross-linking effect of the central tank fuel jettison on the wing tank fuel jettison, but the basic pattern is the same with that of all tanks jettison situation, as shown in Figure 12. Under the given orifice flow area, the rate of weight reduction of the aircraft decreases with the increase of flight altitude.

5. Conclusion

This article proposed an innovative fuel jettison system architecture for a three tank layout wide-body aircraft and the performance has been analyzed. The main conclusion are as following.

(a) The performance of the innovative architecture fuel jettison system is similar to that of the typical architecture system. Besides, the innovative architecture system can improve the safety margin of the fuel system when the wing tanks jettison.

(b) The main pump characteristic and the wing tank jettison orifice flow area have significant impacts on the jettison performance. The overall fuel jettison performance is jointly influenced by the main pump characteristic , override pump characteristic , and fuel line network flow resistance.

(c) The flight altitude has a significant impact on the fuel jettison performance. The fuel jettison performance mainly depends on the fuel pump characteristics at different flight altitudes and the engine fuel flow requirement.

Acknowledgments

The author would like to acknowledge Wang J B, Qu Y Y and Luo J J at Shanghai Aircraft Design and Research Institute for their supports to conduct the test. The author would like to acknowledge Dong L K at Shanghai Aircraft Design and Research Institute for their suggestions to this article.

References

[1] Langton R, Clark C, Hewitt M and Richards L 2009 *Aircraft Fuel System* (Wiltshire: JohnWiley & Sons Ltd.)
[2] SAE International 2019 *Aircraft Fuel System Design Guidelines* AIR7975 (New York: SAE International)
[3] Li M Z 2015 Transport aircraft jettison system design *China Science and Technology Information* **13** pp 62-63
[4] Chen Z B, Wang Y M and Wang J 2015 Study on flight test method for fuel jettison system of civil aircraft *Engineering & Test* **55** pp 48-51
[5] Wang P and Zhu D X 2016 Research on airworthiness requirement of jettison system for large commercial aircraft *Civil Aircraft Design & Research* **3** pp 29-32
[6] Chen G H, Cheng J F, Gu L X and Liu X Q 2019 Simulation and analysis of emergency fuel jettison wake patterns under different flight attitudes *The 4th China Aeronautical Science & Technology Conf.* (Beijing: Chinese Society of Aeronautics and Astronautics) pp 470-476
[7] Yang A J and Wang K 2020 Study on gravity emergency oil discharge of airfoil *Aircraft Design* **40** pp 22-25
[8] Luo J J, Zhu D X, Wang J B, Gong H, Duan A P and Qu Y Y 2021 Fuel jettison test system for aircraft fuel system *CN Patent 202021889212*
[9] Gong H, Zhu D X, Jiang J H, Dong L K and Zhang X Y 2023 System and method for correcting fuel imbalance on the left and right wings of an aircraft *CN Patent 202111247812*

Research on solar sail attitude maneuvering method based on adaptive sliding mode control

Jie Wang[1,*], Tianyi Ma[1], Haibo Song[1], Ye Yu[1]

[1] College of Aerospace Science and Engineering, National University of Defense Technology, Changsha, China

* Corresponding author's email: wangjie@nudt.edu.cn

Abstract. A solar sail is a spacecraft propulsion system that uses the pressure of light to push a spacecraft. It consists of a lightweight support structure that deploys a large area of flexible film. During attitude maneuvering in space, the solar sail will excite vibrations in the flexible support structure and film, which can affect the accuracy of the attitude control system. To address this issue, this article first establishes a rigid-flexible coupled dynamic model of the solar sail that includes attitude motion and flexible structure elastic vibrations. Then, based on an adaptive sliding mode control method, a solar sail attitude maneuvering method is proposed. Finally, a numerical simulation is conducted using a hexagonal-shaped solar sail as the object. The numerical simulation results show that the attitude maneuvering method has the performance of attitude tracking and attitude stabilization, can meet the requirements of large-angle attitude maneuvering for solar sail spacecraft, and has strong robustness to model parameter uncertainty.

1. Introduction

The solar sail accelerates the spacecraft by momentum obtained by reflecting solar photons projected onto the sail surface. Due to the very weak solar pressure generated per unit area, solar sail spacecraft tend to have larger areas and lighter masses. However, the lightweight, ultra-thin construction of the solar sail spacecraft presents new dynamics and poses additional problems.

When a solar sail spacecraft performs orbital or attitude maneuvers in space, it tends to excite nonlinear deformation of the flexible structure, which in turn affects the rigid body motion of the spacecraft. Initially, scholars used the rigid-body model of the solar sail for attitude control system and control strategies using traditional methods such as PID. With the continuous increase in functional demands of spacecraft, the requirements for its attitude control accuracy are also getting higher. It is unable to satisfy the needs of practical missions via traditional control theories. As a result, the control system design used the flexible body model and modern control methods, such as sliding mode control and robust control. Wie[1] used the nonlinear PID control logic for the attitude control of the sailcraft. Actuators for attitude control are two-axis gimbaled boom or vanes[2]. However, the model of the solar sail is a rigid model. Based on the shape variation of booms, Zhang et al.[3] proposed a novel attitude control method, and PD control law was applied to realize the attitude control of three-axis. Eldad et al.[4] established a minimum-time control method of a solar sail with model uncertainty so that the sails were able to perform large angle rotation in a short period of time. Gong et al.[5] proposed an attitude maneuver strategy for the solar sail with axisymmetric spinning. For the control of a flexible 40×40m solar sail, Rotunno et al.[6] used the H_∞, QFT, and input shaping method and the synthesis of four controller types. Firuzi and Gong[7] used the geometrically nonlinear finite element method to study the attitude dynamics and control of a flexible solar sail considering the solar radiation pressure, gravity gradient torque and so on. Furthermore, a robust optimal control approach based on the servomechanism linear quadratic regulator was applied. Although several studies have been carried out in the literature on attitude maneuvers and attitude stabilization of solar sails, in-orbit flight tests of LightSail-2 have shown that the actual control results are significantly different from the design values[8]. Therefore, the high precision attitude stabilization of solar sails is still one of the difficulties of research.

On the other hand, the adaptive sliding mode control method has been widely used for attitude control of spacecraft and has achieved good results. Mostafa et al.[9] broadened the previous applications of variable structure control systems theory to flexible spacecraft's maneuvering. Hu and Ma[10] proposed a robust control algorithm that considered control input nonlinearities. Lo and Chen[11] derived a sliding-mode control (SMC) algorithm, in which the Lyapunov function was utilized to avoid the inverse operation of the inertia matrix. Giri[12] presented a fast finite-time attitude control method using sliding mode strategy. With the bandwidth constraints and unknown disturbances considered, Amrr et al.[13] developed an ET-based adaptive SMC strategy to solve the attitude stabilization problem of spacecraft.

From the above brief review, it is found that the adaptive sliding mode control is effective, and has strong controllability for external interference and internal parameter uncertainty. However, at present, there are few studies on the attitude maneuvering of large-angle of solar sail using sliding mode control. This paper investigates the effectiveness of adaptive sliding mode control methods on rigid-flexible coupled models of solar sails. The structure of this paper is as follows: Section 2 presents the physical configuration of a solar-sail spacecraft; Section 3 presents the rigid-flexible coupling dynamics model of a solar-sail spacecraft; Section 4 proposes a method based on adaptive sliding mode attitude control; Section 5 conducts numerical simulation research; and Section 6 presents conclusions.

2. Problem description

As shown in Figure 1, the solar sail in orbital configuration is a hexagonal shape with a side length of several meters or even tens of meters. The main components of the solar sail include the central body, support booms, and sail surface. The central body is located at the center of the hexagon, with six booms fixed to it. The sail surface consists of six trapezoidal thin films, each tensioned by the central body and booms. At the free ends (i.e., tips) and roots (near the central body) of the trapezoids, the booms connect with the film through a sail-to-structure interface [14].

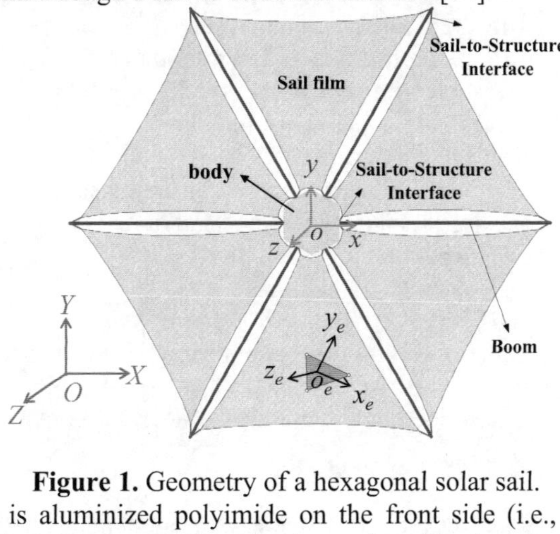

Figure 1. Geometry of a hexagonal solar sail.

The material of the film is aluminized polyimide on the front side (i.e., the sun-facing side). The support booms are lenticular tubes, made of laminating carbon fiber/epoxy prepreg material. Moreover, to ensure that the film remains as flat as possible to maximize the light pressure generated by the sunlight, there is a certain prestress inside the film, which generates a certain pressure on the support booms. However, the lightweight and ultra-thin structural properties bring severe problems to the solar sail spacecraft present. The solar sail film and support booms are so flexible that nonnegligible elastic vibration would be excited when the spacecraft performs attitude maneuvers. And the excited vibration will in turn cause great disturbance to the attitude of the spacecraft. Therefore, it is very important to develop robust attitude control methods to ensure the normal operation of the spacecraft.

3. Rigid-flexible coupling model of solar sail

The solar sail is simplified to a system consisting of a central rigid body, flexible films, and booms. As a result, a hybrid coordinate system is used to describe the movement of the central rigid body and flexible appendage. The Lagrangian method is employed to derive the mathematical model of the system. Firstly, the following coordinate systems need to be defined, as shown in Figure 1. Firstly, the following coordinate systems need to be defined, see. The coordinate system $OXYZ$ is an inertial coordinate system. And the body coordinate system of the spacecraft is represented as $oxyz$. The coordinate system $o_e x_e y_e z_e$ is an element coordinate system used to describe the elastic deformation relative to the central body [15].

Lagrange's equations in terms of quasi-coordinates are shown below [16]

$$\begin{gathered}\frac{d}{dt}\left(\frac{\partial L}{\partial \mathbf{V}}\right)+\tilde{\omega}\left(\frac{\partial L}{\partial \mathbf{V}}\right)=\mathbf{F}_v\\ \frac{d}{dt}\left(\frac{\partial L}{\partial \boldsymbol{\omega}}\right)+\tilde{V}\left(\frac{\partial L}{\partial \mathbf{V}}\right)+\tilde{\omega}\left(\frac{\partial L}{\partial \boldsymbol{\omega}}\right)=\mathbf{M}_\omega\\ \frac{d}{dt}\left(\frac{\partial L}{\partial \dot{\mathbf{q}}}\right)-\left(\frac{\partial L}{\partial \mathbf{q}}\right)=0\end{gathered} \tag{1}$$

Flexible films and booms are discretized using the finite element method. And then Lagrange functions of the system are constructed and the rigid-flexible coupling equation of the solar sail system is derived as

$$\begin{gathered}m\dot{\mathbf{V}}+\mathbf{S}^T\dot{\boldsymbol{\omega}}+\mathbf{C}\ddot{\boldsymbol{\eta}}+\tilde{\boldsymbol{\omega}}(m\mathbf{V}+\mathbf{S}^T\boldsymbol{\omega}+\mathbf{C}\dot{\boldsymbol{\eta}})=\mathbf{F}_v\\ \mathbf{J}\boldsymbol{\omega}+\mathbf{S}\dot{\mathbf{V}}_o+\mathbf{D}\ddot{\boldsymbol{\eta}}+\tilde{\boldsymbol{\omega}}(\mathbf{J}\boldsymbol{\omega}+\mathbf{S}\mathbf{V}+\mathbf{D}\dot{\boldsymbol{\eta}})+\tilde{\mathbf{V}}_o(m\mathbf{V}+\mathbf{S}^T\boldsymbol{\omega}+\mathbf{C}\dot{\boldsymbol{\eta}})=\boldsymbol{\tau}\\ \ddot{\boldsymbol{\eta}}+\xi\dot{\boldsymbol{\eta}}+\boldsymbol{\Lambda}^2\boldsymbol{\eta}+\mathbf{C}^T\dot{\mathbf{V}}+\mathbf{D}^T\dot{\boldsymbol{\omega}}=\mathbf{F}_r\end{gathered} \tag{2}$$

where m is the spacecraft's mass; $\mathbf{J}$ is the spacecraft's inertia matrix; $\mathbf{V}$ is the spacecraft's velocity vector; $\boldsymbol{\omega}$ is the spacecraft's angular velocity vector; $\mathbf{F}_v$ and $\boldsymbol{\tau}$ at the right end of the equation represent the force and moment vector acting on the center body, $\mathbf{F}_r$ is the external force on the flexible attachment, and are expressed in the body coordinate system. $\mathbf{S}$ denotes the coupling matrix of the translation and rotation. $\mathbf{C}$ and $\mathbf{D}$ are the translation and rotation coefficient matrices for rigid-flexible coupling; $\boldsymbol{\eta}(1\times N$, N is the truncation number) is the mode coordinate of the flexible appendage; $\boldsymbol{\xi}=diag[\xi_1, \xi_2, \ldots, \xi_N]$ and $\boldsymbol{\Lambda}=diag[\Lambda_1, \Lambda_2, \ldots, \Lambda_N]$ are the damping ratio matrix and natural frequency matrix for a cantilevered appendage respectively, where ξ_i and Λ_i are the ith damping ratio and frequency [17]. The motion equations of the spacecraft consist of six ordinary differential equations and partial differential equations describing elastic deformation of the connecting appendage. The six ordinary differential equations describe rotation of the rigid body.

4. Sliding mode control method

In this section, an adaptive sliding mode control method for attitude control of solar sail spacecraft is proposed.

The attitude of the spacecraft is described by the modified Rodrigues parameterization

$$\boldsymbol{\sigma}=\hat{\mathbf{e}}\tan\frac{\Phi}{4} \tag{3}$$

where $\boldsymbol{\sigma}=(\sigma_1, \sigma_2, \sigma_3)^T\in\mathbb{R}^3$, $\hat{\mathbf{e}}$ and $\boldsymbol{\Phi}$ are the principal axis and angles of rotation. The kinematics of the rotational speed associated with the angular velocity related to the MRP rate is given as follows

$$\dot{\boldsymbol{\sigma}}=\mathbf{G}(\boldsymbol{\sigma})\boldsymbol{\omega},\quad \boldsymbol{\sigma}(0)=\boldsymbol{\sigma}_0 \tag{4}$$

where

$$\mathbf{G}(\boldsymbol{\sigma})=\frac{1}{2}\left(\mathbf{I}+\tilde{\boldsymbol{\sigma}}+\boldsymbol{\sigma}\boldsymbol{\sigma}^T-\frac{1+\boldsymbol{\sigma}^T\boldsymbol{\sigma}}{2}\mathbf{I}\right) \tag{5}$$

$G(\sigma)$ satisfies

$$\begin{aligned}\boldsymbol{\sigma}^T\mathbf{G}(\boldsymbol{\sigma})\boldsymbol{\omega}&=\left(\frac{1+\boldsymbol{\sigma}^T\boldsymbol{\sigma}}{4}\right)\boldsymbol{\sigma}^T\boldsymbol{\omega}\\ \mathbf{G}^T(\boldsymbol{\sigma})\mathbf{G}(\boldsymbol{\sigma})&=\left(\frac{1+\boldsymbol{\sigma}^T\boldsymbol{\sigma}}{4}\right)^2\mathbf{I}\end{aligned} \tag{6}$$

and $\tilde{\boldsymbol{\sigma}}$ is

$$\tilde{\boldsymbol{\sigma}}=\begin{bmatrix}0 & -\sigma_3 & \sigma_2\\ \sigma_3 & 0 & -\sigma_1\\ -\sigma_2 & \sigma_1 & 0\end{bmatrix} \tag{7}$$

Inverse the matrix $\mathbf{G}(\boldsymbol{\sigma})$ and we can derive

$$\mathbf{G}^{-1}(\boldsymbol{\sigma})=\left(\frac{4}{1+\boldsymbol{\sigma}^T\boldsymbol{\sigma}}\right)^2\mathbf{G}^T(\boldsymbol{\sigma}) \tag{8}$$

Let $\boldsymbol{\sigma}_d$ denotes the desired trajectory, and $\boldsymbol{\omega}_d$ denotes the desired angular velocity. The errors of MRP and angular velocities can be expressed as

$$\begin{aligned}\boldsymbol{\sigma}_e(t)&=\boldsymbol{\sigma}(t)-\boldsymbol{\sigma}_d(t)\\ \boldsymbol{\omega}_e(t)&=\boldsymbol{\omega}(t)-\boldsymbol{\omega}_d(t)\end{aligned} \tag{9}$$

The time-varying surface $\mathbf{s}(t)$ is defined by the errors of MRP and angular velocities

$$\mathbf{s}(t)=\left[\boldsymbol{\omega}(t)-\hat{\boldsymbol{\omega}}(t)\right]+\boldsymbol{\lambda}\mathbf{G}^{-1}(\boldsymbol{\sigma})\boldsymbol{\sigma}_e(t) \tag{10}$$

In the above equation, $\boldsymbol{\lambda}$ is a diagonal matrix with positive elements, and $\hat{\boldsymbol{\omega}}(t)$ is defined as

$$\hat{\boldsymbol{\omega}}(t)=\mathbf{G}^{-1}(\boldsymbol{\sigma})\dot{\boldsymbol{\sigma}}_d(t) \tag{11}$$

After the state moves onto the sliding manifold (s=0), the state would slide along the surface towards $\boldsymbol{\sigma}_d$ and $\boldsymbol{\omega}_d$.
To ensure the sliding manifold is reachable, a feedback control law should be developed.
Choose a Lyapunov function for the closed-loop system as follows

$$V_s=\frac{1}{2}\mathbf{s}^T\mathbf{J}\mathbf{s} \tag{12}$$

Since the inertial matrix $\mathbf{J}$ is a symmetric array and positive definite, V_s satisfies

$$V_s\geq 0 \tag{13}$$

The derivative of this function is

$$\begin{aligned}\dot{V}_s&=\mathbf{s}^T\mathbf{J}\dot{\mathbf{s}}\\ &=\mathbf{s}^T\mathbf{J}\left[\dot{\boldsymbol{\omega}}(t)-\dot{\hat{\boldsymbol{\omega}}}(t)+\boldsymbol{\lambda}\dot{\mathbf{G}}^{-1}(\boldsymbol{\sigma})\boldsymbol{\sigma}_e(t)+\boldsymbol{\lambda}\mathbf{G}^{-1}(\boldsymbol{\sigma})\dot{\boldsymbol{\sigma}}_e(t)\right]\\ &=\mathbf{s}^T[\boldsymbol{\tau}-\boldsymbol{\omega}\times\mathbf{J}\boldsymbol{\omega}-\mathbf{J}\dot{\hat{\boldsymbol{\omega}}}(t)+\boldsymbol{\lambda}\mathbf{J}\dot{\mathbf{G}}^{-1}(\boldsymbol{\sigma})\boldsymbol{\sigma}_e(t)+\boldsymbol{\lambda}\mathbf{J}\mathbf{G}^{-1}(\boldsymbol{\sigma})\dot{\boldsymbol{\sigma}}_e(t)-\boldsymbol{\omega}\times\mathbf{D}(t)\dot{\boldsymbol{\eta}}-\mathbf{D}(t)\ddot{\boldsymbol{\eta}}]\end{aligned} \tag{14}$$

The elastic vibrations of the flexible booms and films induce torques acting on the central body, which can be taken as a disturbance

$$\boldsymbol{\tau}_d=\left[\tau_{d,1},\tau_{d,2},\tau_{d,3}\right]^T=-\boldsymbol{\omega}\times\mathbf{D}(t)\dot{\boldsymbol{\eta}}-\mathbf{D}(t)\ddot{\boldsymbol{\eta}} \tag{15}$$

In a physical spacecraft, the vibrations of the flexible structures are assumed to be limited. As a result, the disturbance can be assumed to be limited and there is

$$\left|\tau_{d,i}\right|\leq\hat{\tau}_{d,i}\qquad(i=1,2,3) \tag{16}$$

where $|\cdot|$ indicates the absolute value of a scalar.
The control law is constructed as

$$\boldsymbol{\tau}=\boldsymbol{\omega}\times\mathbf{J}\boldsymbol{\omega}+\mathbf{J}\dot{\hat{\boldsymbol{\omega}}}(t)-\boldsymbol{\lambda}\mathbf{J}\dot{\mathbf{G}}^{-1}(\boldsymbol{\sigma})\boldsymbol{\sigma}_e(t)-\boldsymbol{\lambda}\mathbf{J}\mathbf{G}^{-1}(\boldsymbol{\sigma})\dot{\boldsymbol{\sigma}}_e(t)+\boldsymbol{\tau}' \tag{17}$$

where

$$\boldsymbol{\tau}' = \left[\tau_1' \ \tau_2' \ \tau_3'\right]^T$$
$$\tau_i' = -k_i \cdot \mathrm{sgn}(s_i), \quad (i = 1,2,3) \tag{18}$$

where sgn(·) is the sign function. Let (17) inserts into (14), and the following formula is induced

$$\begin{aligned}\dot{V}_s &= \mathbf{s}^T(\boldsymbol{\tau}' + \boldsymbol{\tau}_d) \\ &= -\sum_{i=1}^{3} k_i |s_i| \left[1 - \frac{\tau_{d,i}}{k_i} \cdot \frac{s_i}{|s_i|}\right]\end{aligned} \tag{19}$$

By choosing k_i large enough, and k_i should satisfy the inequalities $k_i \geq \hat{\tau}_{d,i}$ (i=1,2,3). Then $\dot{V}_s \leq 0$ can be guaranteed. If $s \neq 0$, $V_s > 0$ and $\dot{V}_s < 0$. As a result, the state of the system moves to the sliding manifold. Therefore, the control law makes the global asymptotic stability of the spacecraft system with an origin error.

The feedback control law in (17) may cause chattering, that is the states are oscillated within a neighborhood of the switching surface. Hence, the sign function is instead by the saturation function $sat(s_i, \epsilon)$ to eliminate chattering,

$$sat(s_i, \epsilon) = \begin{cases} \mathrm{sgn}(s_i), & |s_i| > \epsilon \\ \dfrac{s_i}{\epsilon}, & |s_i| \leq \epsilon \end{cases} \tag{20}$$

5. Numerical simulation

To verify the effectiveness of the adaptive sliding mode control method, numerical simulations are carried out. A solar sail performing a large angle attitude maneuver is taken as the example. Firstly, parameters of the solar sail are given, and then the simulation results are presented.

The parameters of the solar sail are shown in Table 1.

Table 1. Solar Sail Parameters

Component	Properties	Value
Boom	Length (m)	6.67
	Cross-Section area (m^2)	4.3×10^{-5}
	Density (kg/m^3)	1498
	Young's modulus (Gpa)	15.858
	Poisson's ratio	0.2116
	I_x (m^4)	7.08×10^{-9}
	I_y (m^4)	1.74×10^{-8}
Film	Area (m^2)	18.282
	Side length (m)	6.33
	Thickness (mm)	0.0075
	Young's modulus (Gpa)	3.2
	Density (kg/m^3)	1420
	Poisson's ratio	0.37
Cable	Length (mm)	101.7(Inside) 264.2(Outside)
	Area (mm^2)	0.1964
	Young's modulus (Gpa)	65
	Poisson's ratio	0.33
Central body	Central mass (kg)	54
Tip structure	Tip mass (kg)	0.25

The spacecraft performs an attitude maneuver around three axes with an initial moment instanton Euler angle of [0;0;0]. The target value of the attitude is [60;45;30]deg. The corresponded quaternions and MRP are shown in Table 2. The controller parameters are λ = 0.05, k = 0.02, ϵ = 0.002.

The control torque and MRP parameter curves of the spacecraft are shown in **Figure 2** and **Figure 3**. The maximum three-axis control torque during the attitude maneuver is 0.026 Nm. In **Figure 3**, the dashed line indicates the expected value of the MRP parameter, and the solid line is the simulated value of the MRP parameter over time. About 300 seconds from the start of the maneuver, the control moment reaches 0 while the MRP parameters reach their target values.

Table 2. Initial And Target Attitude for Solar Sail

Parameter	Initial Value	Target Value
Euler angles	[0;0;0]	[60;45;30] (deg)
Quaernions	[0;0;0;1]	[0.361;0.440;0.392;0.724]
MRP	[0;0;0]	[0.209;0.255;0.227]
Angular rates	[0.05;-0.02;0.05] (deg/s)	[0;0;0]

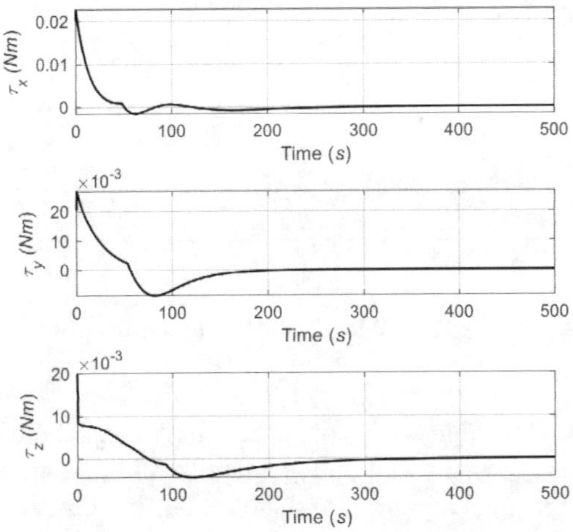

Figure 2. Spacecraft control torque curve.

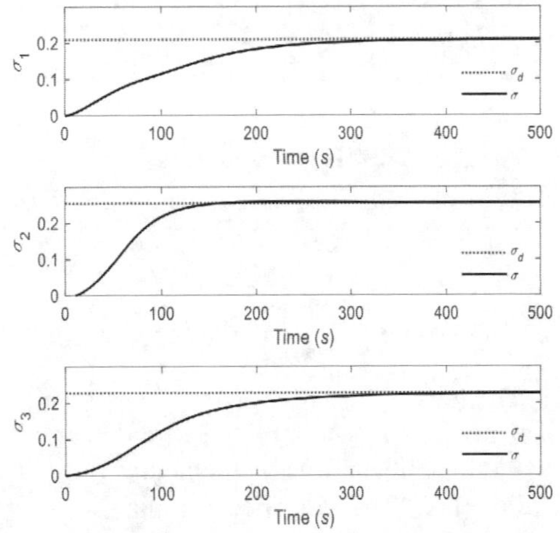

Figure 3. Spacecraft MRP parameter curves.

The response curves of the variables related to the central body attitude are shown in **Figure 4** and **Figure 5**, respectively. As can be seen in **Figure 4**, the spacecraft reached the preset desired attitude after about 300s. The pitch angle showed a small overshoot during the maneuver. In **Figure 5**, there is some deviation in the initial state attitude angular velocity of the spacecraft, which finally converges to 0 after the process of acceleration and then deceleration.

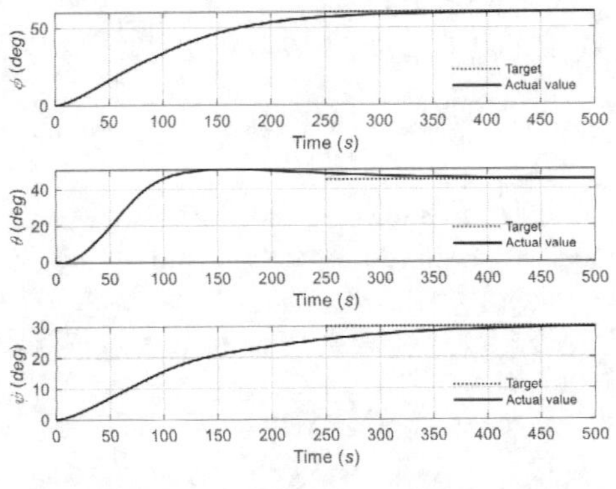

Figure 4. Angular of the central body.

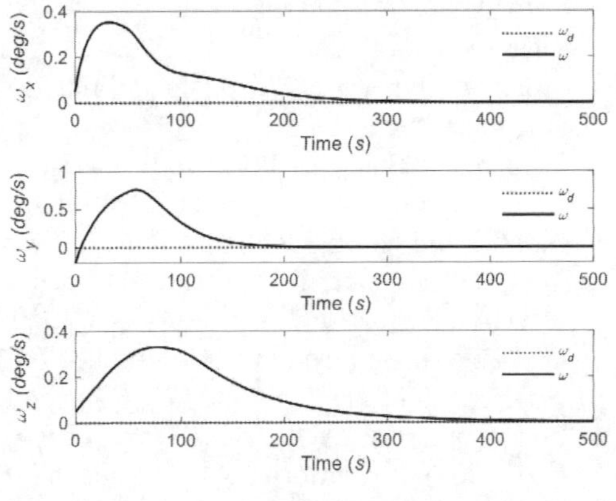

Figure 5. Angular velocity response.

Figure 6 demonstrates the displacement response of the end of the film support boom due to elastic deformation during the maneuver, i.e., the displacement response of the end relative to the center point in the *x*, *y*, and *z*- axis directions of the main body. As can be seen from the figure, the attitude maneuvering of the solar sail excites the out-of-plane vibrations of the support boom. The tip displacement response is mainly determined by the corresponding first-order mode. Meanwhile, due to damping consideration, the vibration of the support boom decays with time. Figure 7 shows the reaction torque generated on the central body by the vibrations of the support boom, with a maximum disturbance torque of 1.3×10^{-4} Nm on the y-axis.

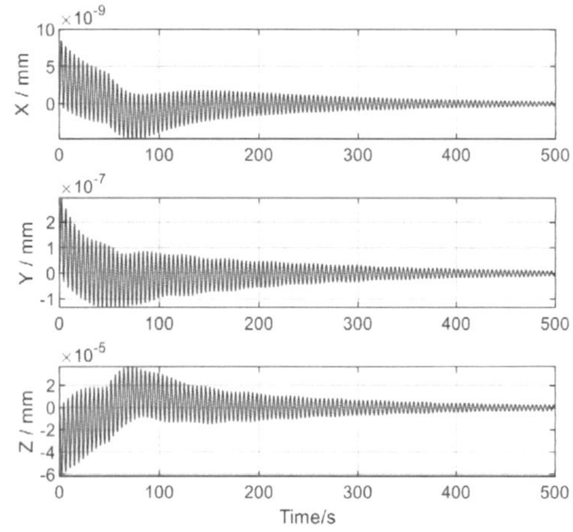

Figure 6. Vibration of the boom tip.

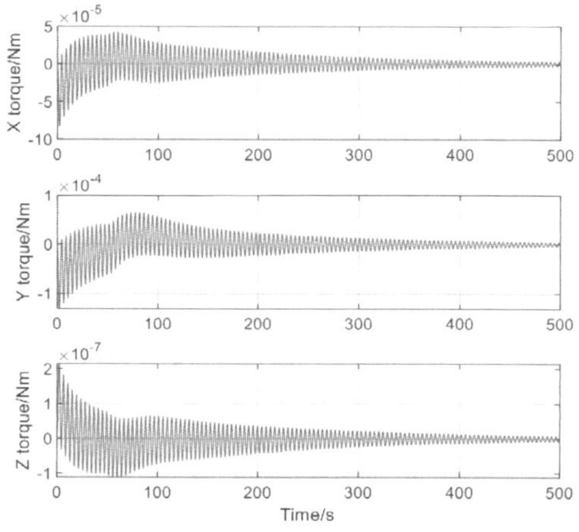

Figure 7. Reaction moment for a support boom.

6. Conclusion

In this paper, the research on the attitude control method of flexible solar sail spacecraft is carried out, and the attitude maneuvering method of the spacecraft is proposed based on the adaptive sliding mode control method. The numerical simulation results using hexagonal solar sails show that the adaptive sliding mode control method can meet the attitude maneuvering and attitude stabilization of the flexible solar sail spacecraft and achieve high precision attitude stabilization under model parameter uncertainty.

References

[1] Wie B 2002 California *AIAA Guidance, Navigation, and Control Conference and Exhibit*, AIAA 2002-4572, pp 1-12

[2] Wie B 2002 California *AIAA Guidance, Navigation, and Control Conference and Exhibit*, AIAA 2002-4573, pp 1-12

[3] Zhang F, Gong S P and Baoyin H X 2021 *Aerospace* Vol **8** No 8 pp 1-21

[4] Eldad O, Lightsey E G and Claudel C 2017 *J. Spacecr. Rockets* Vol **54** No **4** pp 863-70

[5] Gong S P and Li J F 2015 *IEEE Trans. Aerosp. Electron. Syst.* Vol **51** No 2 pp 1462-74

[6] Rotunno M Basso M Pome A B and Sallusti M 2005 California AIAA Guidance, Navigation, and Control Conference and Exhibit AIAA 2005-6083

[7] Firuzi S and Gong S P 2018 *J. Guid. Control. Dyn.* Vol **41** No 8 pp 1715-30

[8] Mansell J, Spencer D A, Plante B, Diaz A, Fernandez M, Bellardo J, Betts B and Nye B 2020 AIAA Scitech 2020 Forum AIAA 2020-2177

[9] OZ H 1986 Astrodynamics Conference AIAA 1986-2193

[10] Hu Q L and Ma G F 2006 *J. Vib. Control.* Vol **12** No 6 pp 659-81

[11] Lo S C and Chen Y P 1995 *J. Guid. Control. Dyn.* Vol **18** No 6 pp 1345-49

[12] Girl D K 2019 California AIAA Scitech 2019 Forum pp 1-21

[13] Amrr S M, Chakravarty A, Alam M M, Algethami A A and Nabi M 2022 *J. Guid. Control. Dyn.* Vol **45** No 7 pp 1328-36

[14] Yu Y, Li D and Wang J 2022 Journal of Physics: Conference Series Vol **2368** No 1 pp 012037

[15] Wang J, Li D 2022 Springer Nature pp 48-49

[16] Wang J, Li D 2018 *Acta Astronautica* Vol **147** pp 393-402

[17] Wang J, Wu J, Liu W and Ji H 2020 4th International Conference on Robotics and Automation Sciences pp 67-71

Reentry trajectory design of a hypersonic vehicle based on reinforcement learning

Partha P Das[1,*], Wang Pei[1] and Chenxi Niu[1]

[1] School of Astronautics, Northwestern Polytechnical University, Xi'an, Shaanxi, China

* Corresponding author's email: partha@mail.nwpu.edu.cn

Abstract. In this research, we investigate control of a hypersonic vehicle (HV) following its reentry into the Earth's atmosphere, using deep reinforcement learning (DRL) in a continuous space. Here, we incorporate the basic kinematic and force equations of motion for a vehicle in an atmospheric flight to formulate the reentry trajectory satisfying the boundary constraints and multiple mission related process constraints. The aerodynamic model of the vehicle emulates the properties of a common aero vehicle (CAV-H), while the atmospheric model of the Earth represents a standard model based on US Standard Atmosphere 1976, with significant simplification to the planetary model. In an unpowered flight, we then control the vehicle's trajectory by perturbing its angle of attack and bank angle to achieve the desired objective, where the control problem is based on different actor-critic frameworks that utilize neural networks (NNs) as function approximators to select and evaluate the control actions in continuous state and action spaces. First, we train the model following each of the methods, that include on-policy proximal policy approximation (PPO) and off-policy twin delayed deterministic policy gradient (TD3). From the trajectory generated, we select a nominal trajectory for each algorithm that satisfies our mission requirements based on the reward model.

1. Introduction

In a hypersonic flight (speed Mach 5 or more), the vehicle that is either returning from space or reentering the planet atmosphere following a low-altitude orbital trajectory, associates a combination of orbital and atmospheric flight mechanics, while needing to maintain a delicate balance between deceleration, heating, and landing accuracy or impact, to ensure safe and reliable navigability of the vehicle [1-2].

Hypersonic reentry of a vehicle in a planetary atmosphere can be classified into either ballistic or lifting entry. In a ballistic reentry, the fundamental restraining force is the drag force, that is directed opposite to the line of flight. In contrast, the lift force which is perpendicular to the flight path acts as the primary decelerating force in case of lifting entry. Manoeuvrability of a lifting vehicle in the atmosphere during hypersonic reentry depends on the lift to drag ratio, which requires variation in angle of attack and possible bank angle change [2-4] which can be achieved in different ways by using propulsion in a powered flight, and by structural dynamics in an unpowered flight by means of reaction control or air deflections from vehicle control surfaces. Key motivation behind this research is to explore control of hypersonic reentry trajectories for manoeuvrable vehicle in an unpowered flight, and develop a possible trajectory design policy by using reinforcement learning (RL) under mission specific constraints, with primary focus being Earth's atmospheric reentry.

RL was demonstrated in an aerospace project [5] for the first time to control an autonomous helicopter flight. Researchers [6] applied deep reinforcement learning (DRL) to command a simulated fleet of wildfire surveillance aircraft. DRL has also been applied for spacecraft orbit control and transfers [7], and to approximate optimal guidance paths for pinpoint planetary landing [8]. Chai et al. [9] presented an integrated reentry trajectory planning and attitude control framework based on deep neural network (DNN). Proximal Policy Optimization (PPO) has been utilized, in hovering and docking with rotating targets [10], in avoidance of on-orbit detection by ground-based sensors [11], and to simulate mid-course exo-atmospheric interception of manoeuvring targets [12]. Miller and Linares, Zavoli and Federici presented PPO's utility for low thrust trajectory optimization in interplanetary missions [13, 14]. Wang et al. demonstrated, feasibility of the DDPG algorithm in autonomous rendezvous mission [15]. Gao et al. used deep reinforcement learning to solve reentry optimization problem of reentry vehicle (RV) based on deep deterministic policy gradient (DDPG) [16]. In a recent work, Hovell and Ulrich proposed a guidance policy for 3-DOF proximity operations using distributed distributional deep deterministic policy gradient (D4PG) [17].

2. Problem formulation

We begin with formulating the reentry dynamics, followed by the definition of necessary mission related constraints including boundary conditions to construct our trajectory design problem.

2.1. Reentry dynamic model

Under the simplifying assumption of a point symmetric spherical Earth rotating with a constant angular velocity, the reentry dynamics equations of motion for a point mass vehicle are [4]

$$\frac{dr}{dt} = V sin\gamma;$$

$$\frac{d\theta}{dt} = \frac{V\, cos\,\gamma\;\, cos\,\psi}{r\, cos\,\phi};$$

$$\frac{d\phi}{dt} = \frac{V\cos\gamma\sin\psi}{r} \tag{1}$$

$$\frac{dV}{dt} = \frac{-D}{m} - g\sin\gamma + \omega^2 r cos\phi(sin\gamma\; cos\phi - cos\gamma\; sin\phi\; sin\psi)$$

$$\frac{d\gamma}{dt} = \frac{L\cos\sigma}{mV} - \frac{g\cos\gamma}{V} + \frac{V\cos\gamma}{r} + 2\omega\cos\phi\cos\psi + \frac{\omega^2 r\; cos\phi(cos\gamma\; cos\phi + sin\gamma\; sin\phi\; sin\psi)}{V}$$

$$\frac{d\psi}{dt} = \frac{L\sin\sigma}{mV\cos\gamma} - \frac{V\cos\gamma\cos\psi\tan\phi}{r} + 2\omega(tan\gamma\; cos\phi\; sin\psi - sin\phi) - \frac{\omega^2 r\sin\phi\cos\phi\cos\psi}{V\cos\gamma}$$

Here, r, θ, ϕ, V, γ, ψ, D, and L define the radial distance, longitude, latitude, velocity, flight path angle, heading angle, drag, and lift force of the vehicle respectively. σ is the bank angle, while m represents the mass of the vehicle. Earth's angular velocity is denoted as ω. Here,

$$L = \frac{\rho V^2 C_L S_{ref}}{2};\; D = \frac{\rho V^2 C_D S_{ref}}{2}$$

where, $C_L = C_{L0} + C_{L1}\alpha + C_{L2}e^{C_{L3}M}$ and $C_D = C_{D0} + C_{D1}\alpha^2 + C_{D2}e^{C_{D3}M}$ are lift and drag coefficients, and the coefficients C_{L0}, C_{L1}, C_{L2}, C_{L3}, C_{D0}, C_{D1}, C_{D2}, and C_{D3} are adopted following the vehicle model of a special class of high performance common aero vehicle (CAV-H) developed by Lockheed-Martin [18]. M is the Mach value of the vehicle.

2.2. Process Constraints

The process constraints during an atmospheric reentry include necessary allowable boundaries for aerodynamic heating, dynamic pressure, and aerodynamic load (normal or g-load) to ensure safety of missions.

2.2.1. Aerodynamic heating. The heating rate $\dot{Q}_s$ is specified generally at the stagnation point on the nose of the vehicle, and depends on the atmospheric density (ρ) and earth relative velocity of the vehicle (V)

$$\dot{Q}(\boldsymbol{x}) = K_Q \rho^{0.5} V^c \leq \dot{Q}_{max} \tag{2}$$

Here, K_Q is the heating rate normalization constant, c usually takes the value of 3 or 3.15, and $\dot{Q}_{max}$ is the maximum allowable heating rate.

2.2.2. Dynamic pressure. Dynamic pressure constraint of the reentry vehicle is represented as

$$P_d(\boldsymbol{x}) = \frac{1}{2}\rho V^2 \leq P_{max} \tag{3}$$

where, P_{max} is the allowable dynamic pressure.

2.2.3. Aerodynamic load. The aerodynamic load (normal load or g-force) is another necessary path constraint, that must be maintained throughout the trajectory. The value for the load constraint is defined as

$$n_L(\boldsymbol{x}) = \frac{\sqrt{L^2+D^2}}{mg_0} \leq N_{max} \tag{4}$$

Here, g_0 is the gravity at mean sea level, L and D are lift and drag force, and N_{max} is the allowable normal load.

2.3. Boundary constraints

We have defined the boundary constraints for the dynamical parameters, by defining each of them (r, θ, ϕ, V, γ, ψ) and control parameters (α, σ) at initial time t_0 , and the terminal values for each at time t_f are defined as following.

$$r(t_0) = r_0;\ \theta(t_0) = \theta_0;\ \phi(t_0) = \phi_0; V(t_0) = V_0; \gamma(t_0) = \gamma_0;\ \psi(t_0) = \psi_0; r(t_f) = r_f;$$
$$\theta(t_f) = \theta_f;\ \phi(t_f) = \phi_f;\ V(t_f) = V_f;\ \gamma(t_f) = \gamma_f;\ \psi(t_f) = \psi_f;\ \alpha(t_0) = \alpha_0;\ \sigma(t_0) = \sigma_0 \tag{5}$$

The values for all this parameter are generally set in a bounded domain for each category as,

$$r_{min} \leq r \leq r_{max};\ \theta_{min} \leq \theta \leq \theta_{max}; \phi_{min} \leq \phi \leq \phi_{max};$$
$$V_{min} \leq V \leq V_{max};\ \gamma_{min} \leq \gamma \leq \gamma_{max};\ \psi_{min} \leq \psi \leq \psi_{max};$$
$$\alpha_{min} \leq \alpha \leq \alpha_{max};\ \sigma_{min} \leq \sigma \leq \sigma_{max};\ |\dot{\alpha}| \leq \dot{\alpha}_{max};\ |\dot{\sigma}| \leq \dot{\sigma}_{max} \tag{6}$$

The min and max values for each parameter denote minimum and maximum allowable range for the states and controls. $\dot{\alpha}$ and $\dot{\sigma}$ are the rate of changes for angle of attack and bank angle respectively, and bounded by maximum rate of change in both as $\dot{\alpha}_{max}$ and $\dot{\sigma}_{max}$.

3. Research methodology

In this section, we have presented the implementation method, that we have adopted for a reentry trajectory control framework of a hypersonic vehicle based on reinforcement learning. A reinforcement learning agent (learner or decision maker) and its environment (everything outside the agent) interact over a sequence of discrete time steps. Under the Markov decision process (MDP), we have constructed the environment by defining the state, action and reward model to compute the trajectory of a chosen reentry vehicle (RV). We have adopted different on/off-policy actor-critic frameworks based on the policy gradient theorem, particularly suitable for continuous action and state space. Utilizing neural networks as function approximators, on-policy proximal policy optimization (PPO) [19] and off-policy twin delayed deep deterministic policy gradient (TD3) [20] methods have been applied to train and evaluate the model, applicable and nominal in fulfilling certain mission requirements. The objective is to find a nominal trajectory satisfying

$$\min(d_h) = |r_{goal} - r(t_f)|;\ \min(d_\theta) = |\theta_{goal} - \theta(t_f)|;$$
$$\min(d_\phi) = |\phi_{goal} - \phi(t_f)|;\ \min(d_v) = |V_{goal} - V(t_f)|; \tag{7}$$

in optimal time,

$$\min\left(\int_{t_0}^{t_f} dt\right) = \min(t_f - t_0) = t^* \tag{8}$$

Here, d_h represents the difference between objective/target radial distance r_{goal} and terminal radial distance $r(t_f)$, d_v represents difference between objective/target velocity V_{goal} and terminal velocity $V(t_f)$, d_θ represents difference between objective/target longitude θ_{goal} and terminal longitude $\theta(t_f)$, and d_ϕ stands for the difference between objective/target latitude ϕ_{goal} and terminal latitude $\phi(t_f)$.

3.1. RL environment

One of the two core simulation frameworks implemented in our study for trajectory control is built on the definition of state, action, and reward model, that the other core consisting a compact actor-critic RL architecture can interact with.

3.1.1. State space. In our study, the environmental state at time t, denoted as S_t, includes the radial distance (r), longitude (θ), latitude (ϕ), velocity (V), flight path angle (γ), and the heading angle (ψ) of the RV. Future states S_{t+1} are derived from the vehicle dynamics by numerically integrating the set of kinematic and force equations of motion specified in equation (1) using Euler method. Hence,

$$S_t = [r, \theta, \phi, V, \gamma, \psi] \tag{9}$$

3.1.2. Action space. Conventionally, in an unpowered hypersonic reentry flight, control of a vehicle is achieved by actions taken by the agent in the form of aerodynamic deflections, depending on the perturbation of angle of attack (α) and bank angle (σ), to follow a desired trajectory. The action A_t at time t, can be written in vector form as

$$A_t = [\alpha, \sigma] \tag{10}$$

3.1.3. Reward function. The reward signal consists of a primary performance reward (R_p) based on the differences in current and terminal states, angular direction reward/penalty (R_d), sparse trajectory corridor reward/penalty (R_s), terminal reward/penalty, and some additional performance bonuses (b). Following are the general representations of reward function attributes mentioned

$$R_p = \begin{cases} \varepsilon\begin{pmatrix} \kappa - |r(t) - r_{goal}|/c_h - |V(t) - V_{goal}|/c_v \\ -|\theta(t) - \theta_{goal}| \times c_\theta - |\phi(t) - \phi_{goal}| \times c_\phi \end{pmatrix} & \text{if not terminated} \\ -\chi & \text{if terminated} \end{cases} \tag{11}$$

Here, ε, κ, c_h, c_v, c_θ, c_ϕ, and χ are real valued constants, having adjusted values for different algorithms applied, while $|r(t) - r_{goal}|$, $|v(t) - V_{goal}|$, $|\theta(t) - \theta_{goal}|$, and $|\phi(t) - \phi_{goal}|$ represent absolute differences in current state and terminal/objective state for radial distance, longitude, latitude and velocity respectively.

$$R_d = \begin{cases} \ln(n) & \text{if} \quad |\measuredangle_r - \measuredangle_o| < 0.000001 \\ \ln(n) \times 0.2 & \text{if } \measuredangle_o \times 0.99 \le \measuredangle_r \le \measuredangle_o \times 1.01 \\ \ln(n) \times 0.1 & \text{if } \measuredangle_o \times 0.98 \le \measuredangle_r \le \measuredangle_o \times 1.02 \\ -\ln(n) \times 0.1 & \text{otherwise} \end{cases} \tag{12}$$

$\measuredangle_r$ and $\measuredangle_o$ are real and directional objective angles respectively, and n is number of steps.

$$R_s = \begin{cases} n \times 0.5 \\ \text{if } \left\{ n \times \left(\frac{\theta_{diff}}{10}\right) \leq \theta(t) \leq (n+1) \times \left(\frac{\theta_{diff}}{10}\right) \right\} \\ \cap \left\{ n \times \left(\frac{\theta_{diff}}{10}\right) \leq \theta(t) \leq (n+1) \times \left(\frac{\theta_{diff}}{10}\right) \right\} \\ -0.5 \qquad \text{otherwise} \end{cases} \tag{13}$$

where, $n = 0, 1, \ldots, 9$ and $\theta_{diff} = \theta_{goal} - \phi_0$ is the difference in longitude between initial and objective values.

Hence, total reward per episode stands as,

$$R_T = \sum(R_p + R_d + R_s) \tag{14}$$

4. Results and discussion

In our effort to investigate unpowered trajectory control of a reentry hypersonic vehicle in the Earth's planetary atmosphere, we have applied reward based actor-critic reinforcement learning methods to approximate the desired policy that returns the best reward for some certain preconditions. Table 1 shows designed scenario with initial values for each state and action parameters, with primary objective to reach the positional target of longitude, latitude, and radial distance (velocity, flight path angle, and heading objectives can be considered as soft goals). A non-thrusting control action achieved through the perturbation of angle of attack and banking angle is assumed and adopted along with the necessary system credentials.

Table 1. Mission scenario with initial states and action values along with objective states, and state and action space minimum and maximum values.

Dynamic Parameters	Initial values s_0, a_0	Terminal values s_f, a_f	s_{min}, a_{min}	s_{max}, a_{max}
Radial Distance, r (km)	6470	6400	6378	6488
Longitude, θ (°)	0	70	-180	180
Latitude, ϕ (°)	0	40	-90	90
Velocity, V (m/s)	7000	1000	7050	800
Flight Path Angle, γ (°)	-2	-5	-90	90
Heading Angle, ψ (°)	45	0	-180	180
Angle of Attack, α (°)	25	n/a	5	30
Bank Angle, σ (°)	30	n/a	-90	90

The agents are trained for a total of 1 million timesteps each, and each path taken by the vehicle has been analyzed based on their average or total return. The reward model discussed in the previous chapter has been utilized to form a nominal trajectory. In table 2 process constraints considered are given, followed by the related necessary constants in table 3.

Table 2. Process/ aerodynamic constraints.

Maximum Aerodynamic Heating Rate, $\dot{Q}_{max}$	Maximum Dynamic Pressure, P_{max}	Maximum Aerodynamic Load, N_{max}
200 (kW/m$^{2)}$	500 (kPa)	10 (g)

Table 3. Default constants.

Constant	Value	Constant	Value
Earth Radius (km)	6378	Earth Gravitational Parameter, μ	3.986×10^{14}
Earth Angular Speed (rad/s)	7.292115×10^{-5}	Aerodynamic Heating Constant	5.188×10^{-8}

To train the model we have adopted the relevant hyperparameters and neural network architecture for each actor-critic method according to the original papers cited earlier. Each training session has returned some policies, from which we have chosen the nominal trajectories, that have returned maximum reward (Maximum average rewards were also similar for both) from the PPO and TD3 algorithmic framework. The outcome from the reproduction of the selected trained policies are summarized below in table 4 and successive plots. The range to target is calculated according to the haversine formula used to evaluate the great circle distance.

Table 4. Terminal values of all the six states and final cross-range distance.

Parameters	Objective	PPO	TD3
Radial distance, r (km)	6400	6402.9114	6406.3210
Longitude, θ (°)	70	70.0089	69.9221
Latitude, ϕ (°)	40	40.0718	40.1275
Velocity, V (m/s)	1000	2603.8830	935.1989
Flight Path angle, γ (°)	-5	-2.1429	-3.8657
Heading angle, ψ (°)	0	5.6542	13.7927
Range to target, R (km)	< 15	8.0598	15.7375
Duration of flight, t (s)	n/a	1641.5	1785.5

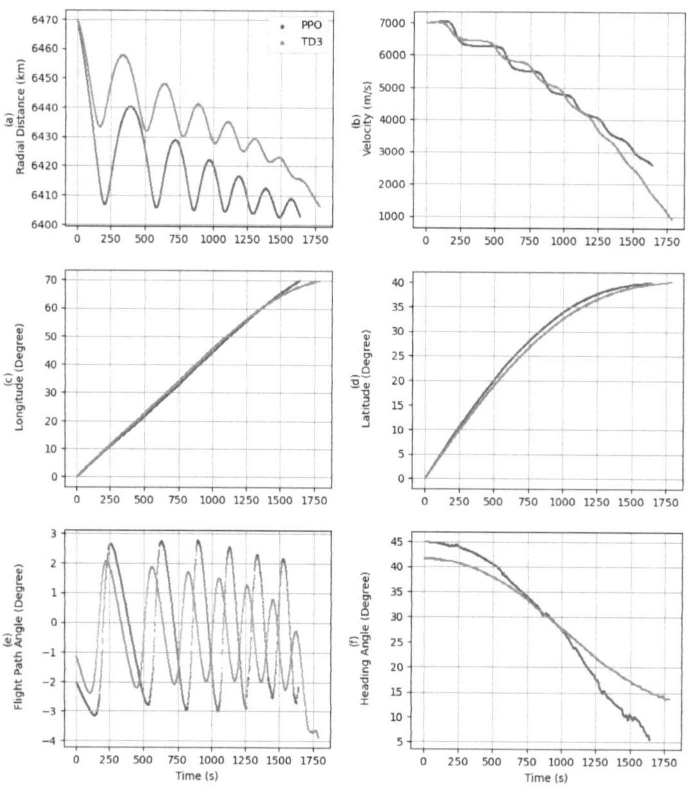

Figure 1. Performance of the selected PPO and TD3 trained policies showing radial distance [km] (a), velocity [m/s] (b), longitude [°] (c), latitude [°] (d), flight path angle [°] (e), & heading angle [°] (f).

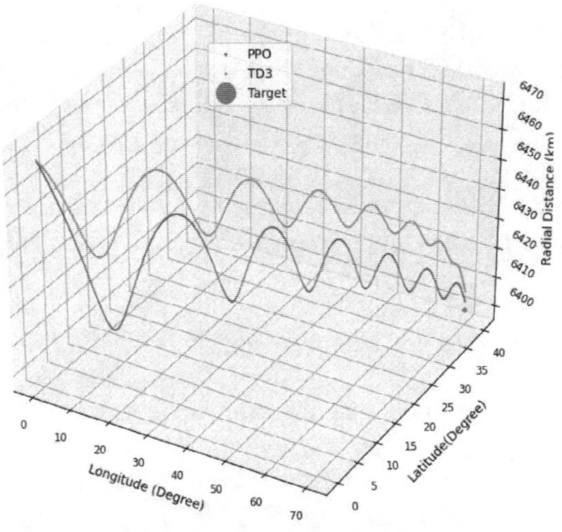

Figure 2. Three dimensional trajectories for the selected PPO and TD3 policies.

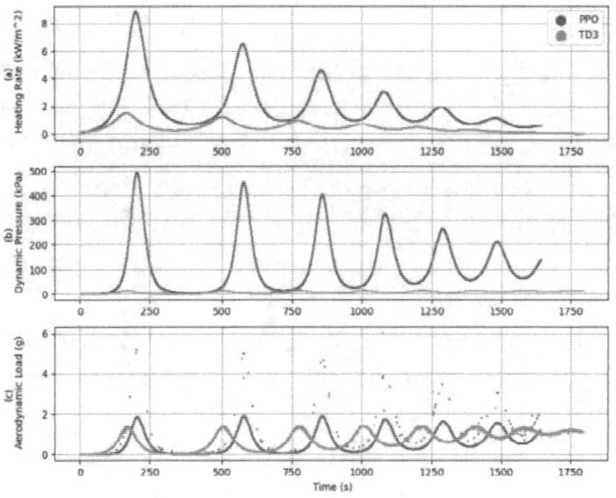

Figure 3. Aerodynamic Heating Rate (a), Dynamic Pressure (b), and Aerodynamic Load (c) for the selected trajectories.

In the above graphs, first we have presented performances of the PPO and TD3 trained policies depicting all the six states during the flight (figure 1), followed by corresponding three-dimensional trajectory representation in achieving our primary objective of reaching the target position (figure 2). The aerodynamic constraints measures are demonstrated in figure 3. These results (Table 4 and Figure 1-3) clearly show the PPO trajectory has reached the closest to the target in terms of both altitude and range to the target. The trajectory taken by TD3 however has come closest to one of our key soft goals of terminal velocity and also has maintained the safest aerodynamic heating, pressure, and load values.

5. Conclusion

During our study, we have found some key features of implementing RL in a reentry scenario which include but are not limited to: selection of suitable initial conditions for desired terminal condition is paramount, tuning of reward model accordingly may require extensive trial and error processes, and generalization of the policies meaning their applicability in an unknown scenario is not viable yet. Reinforcement learning offers a magnitude of opportunities for trajectory control of HV in planetary reentry environment with extensive training, given well defined state and action boundaries. Further researches should be conducted to better understand the mechanism of RL algorithm and unlock the full potential of it in hypersonic trajectory control.

References

[1] Chapman D R 1960 *An Analysis of the Corridor and Guidance Requirements for Supercircular Entry into Planetary Atmospheres*. Washington, DC: U.S. Gov. Print. Off..
[2] Adams Jr J C 2003 Atmospheric Re-Entry *Exoaviation.*
[3] Dickmanns E D 1972 Maximum range three-dimensional lifting planetary entry National Aeronautics and Space Administration Washington D. C. USA Tech. Rep. NASA TR R-387.
[4] Vinh N X 1981 *Optimal trajectories in atmospheric flight* New York: Elsevier.
[5] Abbeel P, Coates A, Quigley M and Ng A. Y. 2006 An application of reinforcement learning to aerobatic helicopter flight *Advances in neural information processing systems 19.*
[6] Julian K and Kochenderfer M 2019 Distributed wildfire surveillance with autonomous aircraft using deep reinforcement learning *Journal of Guidance, Control, and Dynamics* vol. 42 no. 8 pp. 1768-1778.
[7] LaFarge N, Miller D, Howell K and Linares R 2020 Guidance for closed-loop transfers using reinforcement learning with application to libration point orbits *AIAA Scitech 2020 Forum.*

[8] Gaudet B, Linares R and Furfaro R 2020 Deep Reinforcement Learning for six degree-of-freedom planetary landing *Advances in Space Research* vol. 65 no. 7 pp. 1723–1741.
[9] Chai R, Tsourdos A, Savvaris A, Chai S, Xia Y and Chen C L P 2020 Six-DOF spacecraft optimal trajectory planning and real-time attitude control: a deep neural network-based approach *IEEE Transactions on Neural Networks and Learning Systems* vol. 31 no. 11 pp. 5005-5013.
[10] Oestreich C E, Linares R and Gondhalekar R 2021 Autonomous Six-degree-of-freedom spacecraft docking with rotating targets via reinforcement learning *Journal of Aerospace Information Systems* pp. 1–12.
[11] Reiter J A, Spencer D B and Linares R 2019 Spacecraft detection avoidance maneuver optimization using reinforcement learning *29th AAS/AIAA Space Flight Mechanics Meeting* pp. 3055–3069, Univelt Inc..
[12] Gaudet B, Furfaro R and Linares R 2020 Reinforcement learning for angle-only intercept guidance of maneuvering targets *Aerospace Science and Technology* vol. 99 p. 105746.
[13] Miller D and Linares R 2019 Low-thrust optimal control via reinforcement learning," *Advances in the Astronautical Sciences* Vol. 168 pp. 1817–1834.
[14] Zavoli A and Federici L 2020 Reinforcement learning for low-thrust trajectory design of interplanetary missions *2020 AAS/AIAA Astrodynamics Specialist Virtual Lake Tahoe Conference.*
[15] Wang X, Wang G, Chen Y and Xie Y 2020 Autonomous rendezvous guidance via Deep Reinforcement Learning *2020 Chinese Control And Decision Conference (CCDC).*
[16] Gao J *et al.* 2019 Reentry trajectory optimization based on Deep Reinforcement Learning *2019 Chinese Control And Decision Conference (CCDC)* pp. 2588-2592.
[17] Hovell K and Ulrich S 2021 Deep reinforcement learning for spacecraft proximity operations guidance *Journal of Spacecraft and Rockets* vol. 58 no. 2 pp. 254-264.
[18] Song J, Tong X, Xu X and Zhao K 2022 A real-time reentry guidance method for hypersonic vehicles based on a time2vec and Transformer Network *Aerospace* vol. 9 no. 8 p. 427 doi:10.3390/aerospace9080427.
[19] Schulman J, Wolski F, Dhariwal P, Radford A and Klimov O 2017 Proximal policy optimization algorithms *arXiv preprint arXiv:1707.06347.*
[20] Fujimoto S, Hoof H and Meger D 2018 Addressing function approximation error in actor-critic methods *International conference on machine learning* pp. 1587-1596.

Research on safety issues and safety analysis of new energy aircraft

Tang Chu[1,*], **Li Zhongyang**[2], **Wang Libo**[2]

[1]Unmanned System Research Institute, Northwestern Polytechnical University, China
[2]AVIC the First Aircraft Institute, China

* Corresponding author's email: tc@nwpu.edu.cn

Abstract. As the global aviation industry faces increasing pressure on the environment, the International Civil Aviation Organization (ICAO) has implemented increasingly strict requirements for aircraft carbon emissions. The adoption of innovative energy technologies, such as electric power, hydrogen fuel, and sustainable biofuels in the aviation sector, will enable us to achieve zero carbon emission targets and mitigate environmental pollution associated with air transportation. Drawing on statistical data of aviation accidents, this paper analyses the primary causes of flight mishaps and summarizes several safety issues faced by new energy aircraft. Based on these, the new energy aircraft is divided into several subsystems, including power system, energy system, flight control system and so on. Then, safety analysis tools and methods are then used to conduct an analysis of the aircraft's safety, with the aim of improving overall safety.

1. Introduction

With the aviation industry continuously expanding, environmental and energy issues are becoming increasingly salient. The future's strict environmental standards present a challenge for traditional fuel aircraft. The use of clean energy to gradually replace fuel power has become an important trend in the development of the aviation industry.

In recent years, there has been a global surge in the development of new energy aircraft, like the revolution seen in new energy cars. As the aviation industry continues to expand, environmental and energy issues are becoming increasingly prominent [1]. The technology for developing and producing new energy aircraft is advancing rapidly. Airbus, Boeing, Dassault, and other airlines have unveiled new energy concepts. Rolls-Royce has recently successfully tested hydrogen-powered aircraft engines. FAA, EASA, and CAAC have successively established committees that are specifically responsible for setting airworthiness standards for energy aircraft. Additionally, regulations and standards have been approved for these types of aircraft [2, 3]. New energy aircraft are expected to enter the commercial market within the next 5-10 years.

Thanks to the strong promotion of green aviation development, new energy aircraft can develop rapidly. At present, enterprises and research institutions worldwide are primarily focused on technical demonstrations and designs. The battery energy density limits the feasibility of pure electric aircraft for both trunk and regional lines. However, hydrogen fuel technology has been identified as a viable alternative for constructing such lines in other countries, although it is still in its early stages. Hydrogen fuel-powered aircraft is a medium- and long-term development way for new energy aircraft. By around 2050, there will be a complete line of fully autonomous new energy aircraft industry products that cover all markets and product lines [4-6]. Many of the system technologies lack established standards and norms for airworthiness certification and safety. It is necessary to address the safety issues and key technologies of new energy aircraft.

2. Analysis of accident data

Aviation accident data from 1950 to 2019, for which the cause has been determined, is analysed for commercial aircraft carrying at least 19 passengers. The pie chart displays five categories of accident causes in figure 1, including human error, engine failure, weather conditions, artificial damage, and other. Human error was identified as the primary cause of failure, accounting for 49% of aircraft accidents. Engine failures accounted for 23%, following human errors. Given that aviation accidents typically result from multiple factors, the primary or initiating cause is often identified. Environmental factors were found to be the third most common contributing factor to accidents.

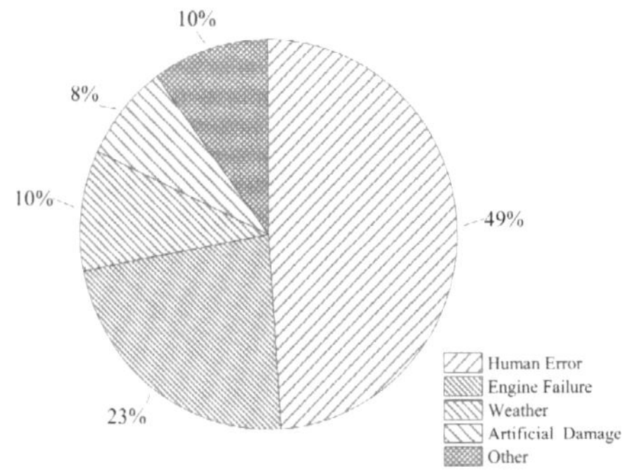

Figure 1. Failure modes observed in commercial aircraft accidents between 1950 and 2019.

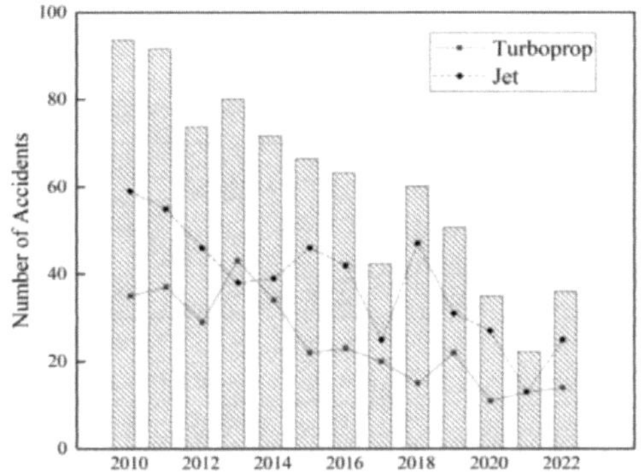

Figure 2. The accidents data compiled by IATA for 2010 to 2022.

Figure 2 reviews the accident data compiled by the International Air Transport Association (IATA) for 2010 to 2022. The results indicate that the safety of IOSA-certified airlines has improved to some extent over the past decade, as evidenced by a decreasing number of accidents each year. The number of major damage accidents involving civil jet aircraft is higher than that of turboprop aircraft, and these accidents most frequently occur during the landing and low-altitude flight stages. The safety of turboprop aircraft has not significantly improved over the past five years.

3. Safety issues for new energy aircraft

The primary features of new energy aircraft are their environmental friendliness and economic viability. The main difference between new and traditional aircraft lies in the utilization of alternative energy sources. The use of new energy sources, such as solar power, electric power, and hydrogen fuel, has resulted in modifications to the power plant and corresponding advancements in aircraft aerodynamics, energy systems, power systems, and flight controls. Modifications made at the design level will have an impact on security concerns.

3.1. Energy system safety issues

The energy system of an aircraft can be classified into primary and secondary energy systems [7]. The primary energy system of new energy aircraft converts clean sources of energy, such as electricity, hydrogen, and solar power, into the driving force for propulsion. The current focus of new energy aircraft is primarily on small electric aircraft. However, the limited energy density of storage equipment still presents challenges for trunk and regional aircraft, which are still in the conceptual stage. Storing hydrogen fuel in large quantities and safely on airplanes is a challenge. High-pressure gas storage occupies more space in the airframe, while cryogenic liquid storage offers higher energy density but requires demanding conditions.

Additionally, high concentrations of hydrogen can easily cause combustion due to its invisible and heatless nature, making it difficult to detect. All these factors have led to serious flight safety accidents. Therefore, it is necessary to implement effective monitoring and protective measures, as well as establish corresponding safety standards for the use and monitoring of hydrogen. For new energy aircraft, the configuration of the energy system differs, requiring a specific focus on safety analysis and tailored problem-specific analyses.

3.2. Electric propulsion system safety issues

The conversion of secondary energy systems from traditional aircraft into electric energy signifies progress towards the electrification of aircraft energy, which is a crucial aspect in the development of sustainable aviation. Electric propulsion technology represents a significant innovation in energy and power systems, marking an advanced stage in the development of aircraft electrification [8]. The research on new energy aircraft primarily focuses on electric propulsion units and hybrid power systems. This type of power plant needs to pay attention to electrical hazards and ensure effective measures for high voltage and fire protection [9, 10]. The American Society for Materials has formulated the ASTM F2840 standard, which sets forth requirements on airworthiness safety issues such as the use of materials and components in electric propulsion devices, establishment of operational limits, environmental conditions, simulation calculations, and airworthiness verification.

3.3. Flight control system safety issues

With the advancement of computer technology, the entire aircraft system management platform has been optimized for autonomous control and improved flight performance through real-time monitoring of flight status [11]. Flight control system technology is developing towards automation, autonomy, greater efficiency and safety, in order to meet the challenges of integrating new energy aircraft.

Therefore, the control system of new energy aircraft must meet the following requirements: a) The flight control computer should have robust computational capabilities and a sophisticated monitoring program to cater to the demands of intricate integrated systems. b) The design of complex and highly integrated systems must have sufficient margins and comply with engineering reliability and airworthiness safety standards, and it is crucial to explore additional methods for conformity verification. c) the Important subsystems and components should be backed up and easily switched to the standby operating system in case of primary system failure.

3.4. Energy storage system security issues

The energy storage systems utilized in new energy aircraft primarily include batteries, fuel cells, solar cells, supercapacitors, liquid hydrogen tanks, and biofuel tanks. The lithium-ion batteries, serving as representatives of advanced energy storage systems, possess immense potential and are regarded as the fundamental components of new energy electric vehicles. Based on statistics from 2014 to 2019 regarding electric vehicle explosions, the primary causes include spontaneous ignition, collisions accident, equipment failure, battery charging, cable aging, short circuits, immersion events, human operation and so on, as figure 3. These factors can lead to thermal runaway and ultimately result in an explosion with the battery being identified as the main culprit.

The primary factors contributing to energy storage system accidents can be categorized as follows: a) Battery system deficiencies; b) Inadequate protection systems; and c) Complex operational environments.

3.5. Composite material safety issues

The extensive use of advanced composite materials in new energy aircraft is an inevitable trend for future development. China has conducted years of research and development on aviation composite materials, leading to the gradual establishment of a comprehensive technical system that essentially meets the requirements of relevant technical standards for aerospace applications. However, ensuring the structural integrity, damage detection capabilities, and proper material maintenance procedures of composite materials is crucial for upholding aircraft reliability and airworthiness.

Journal of Physics: Conference Series **2633** (2023) 012006 doi:10.1088/1742-6596/2633/1/012006

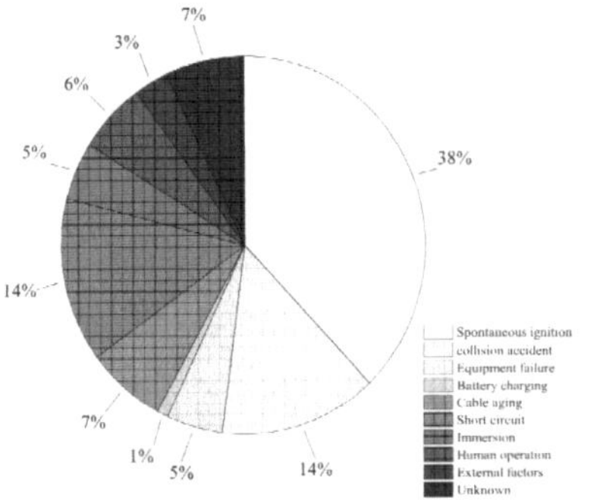

Figure 3. Statistics on electric vehicle explosion accidents from 2014 to 2019.

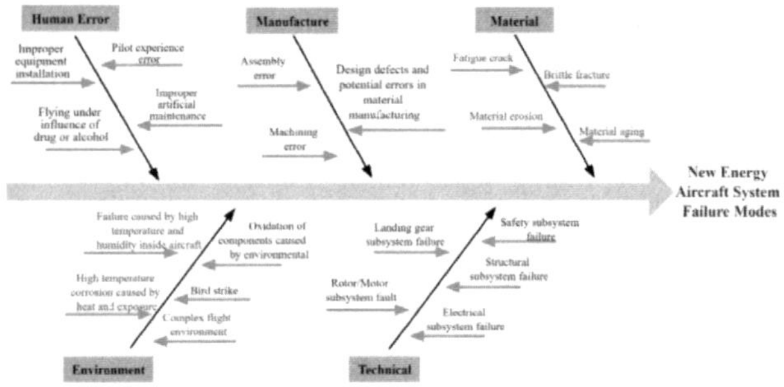

Figure 4. New Energy Aircraft System Fishbone Diagram.

3.6. Aerodynamic configuration safety issues

The integration of aerodynamic structure and propulsion design is an important trend in the advancement of new energy aircraft, as it endeavours to augment aerodynamic performance [12, 13]. The primary design and manufacturing directions for new energy trunk and regional aircraft are distributed electric propulsion and wing-body configuration. This innovative arrangement provides structural, material, and power advantages for the aircraft. However, due to a lack of suitable analytical tools and references, ensuring airworthiness safety remains challenging.

4. Safety analysis method

The safety of new energy aircraft is assessed through the application of a safety analysis methodology, which draws on statistical analyses of flight accidents and their associated safety concerns. For example, for hydrogen-fueled aircraft, it is necessary to pay attention to the safety problems, such as Hydrogen storage tank and Hydrogen storage line, and for pure electric aircraft, electric energy is used for all or part of the propulsion system. Components requiring safety considerations include batteries, motors, cables and related equipment.

4.1. Function hazard analysis (FHA)

According to the SAE ARP4961 recommended process for assessing aircraft system safety, the FHA scope extends throughout the development life cycle. For the purposes of the Preliminary System Safety Assessment (PSSA), the FHA considers only functional failure modes (i.e., not their probability of occurrence) [14, 15]. Considering the difference between new energy aircraft and traditional aircraft, it is imperative to comprehensively identify the functions of the new energy aircraft system and its subsystems, account for potential functional failures as well as combined failures across systems, and assess their impact levels.

Table 1. FHA table.

Function	Failure state	Failure phase	Impact of failure on aircraft or personnel	Failures and other impacts	Impact level	Compliance verification method	Remarks
		•Taxing •Take off •Cruise •Descending •Landing •Approach		•For plane •For crew •For passengers	•I: disastrous impact; •II: serious impact •III: significant impact •V: smaller impact	Quantitative FMEA Analysis/Quantitative FTA Analysis /CCA	

The safety requirements identified through aircraft-level FHA and system-level FHA can be addressed by utilizing Fault Tree Analysis. The purpose of using FHA is to determine the top event and hazard level associated with a specific risk. The FHA table typically includes the main contents shown in the following table 1.

4.2. Failure modes and effects analysis (FMEA)

The FMEA is a structured way of identifying actions that could lead to collapse of a system, subsystem, or component. Its aim is to evaluate the consequences resulting from hypothetical failures of components within the system. The new FMEA standard, which was published in 2019, enhances the implementation process and introduces the concept of Seven-step Method (as table 2). By conducting quantitative analysis of the risk priority number (RPN), it is possible to identify the fault mode posing the highest risk to the system. This risk analysis is achieved by combining it with the action priority (AP) method.

Table 2. Seven-step Method.

Systems Analysis			Failure Analysis/Risk Reduction			Spacing
1st step	2nd step	3rd step	4th step	5th step	6th step	7th step
Planning and preparation	Frame analysis	Functional analysis	Failure analysis	Risk analysis	Optimization	Result documentation

Utilizing the research of new energy aircraft system, an FMEA was completed by decomposing the overall system into five subsystems: power system, propulsion system, flight control system, electrical system, and structural system. While each subsystem has technical failure modes only applicable to its respective area, there are noticeable failure modes that repeat amongst each subsystem: manufacturing, human error, material fatigue, and environmental. In accounting for these redundant failures, a separate category for overall system failures was organized where these common subsystem failures were explored separately from the unique technical failure modes. The failure of each subsystem is standardized based on its severity (S), occurrence (O), and detection (D). Consequently, the risk coefficient RNP is calculated by multiplying S, O, and D.

Table 3. RNPs from the FMEA.

Subsystem	Categories	Fault Mode	Potential impact of failure	Severity (S)	Occurrence (O)	Detection (D)	AP
				Influence/value			
Power system	Engine	Blade damage Drive shaft bearing failure Gear failure	Operational overloads of the motor components, premature wear	Very high (8-10) High (6-7) Middle (4-5) Low (2-3)			H: (Appropriate measures need to be identified to improve prevention efforts) M: (Appropriate measures should be
Landing Gear system	Extender	Landing gear retraction and release Direction control Deceleration	Structural damage				
Flight con-	Equipment	The sensor is faulty and the in-					

Journal of Physics: Conference Series **2633** (2023) 012006 doi:10.1088/1742-6596/2633/1/012006

trol system		put signal is lost Flight control computer malfunctioned		Very low (1)	identified to improve prevention)
Electrical Subsystem	Battery	Battery protection	Charge builds up or overheating		L: (Measures can be identified to improve prevention)
		Battery leakage	Defective battery, poor temperature control and aging		
Propulsion systems	Propulsion	Failure of mechanisms	Lack of forward thrust forces leading to catastrophe		

This resulted in 4-5 components or functions for cause of failure within a subsystem and 25 overall components to be analysed into potential failure modes and potential causes and effects of these failures. Then, the failure modes requiring action are classified by priority to reduce the magnitude of risk. Here, AP can be categorized into high (H), medium (M), and low (L) priority levels.
Table 3 shows the RPN failure modes.

4.3. Fishbone diagram

From the information gathered in the FMEA, Fishbone Diagrams based on five subsystem failures were obtained that can be used to study overall system failures. Such graphical diagrams provide a clear depiction of the potential causes of the FMEA and clearly show the major root causes of these subsystems, making it easier to understand how these subsystems work and affect the whole-aircraft system of the new energy aircraft. figure4 shows the overall system Fishbone Diagram. Which looks at the redundant failure modes that affect all subsystems, as well as the technical aspects of each subsystem.

4.4. Fault tree analysis (FTA)

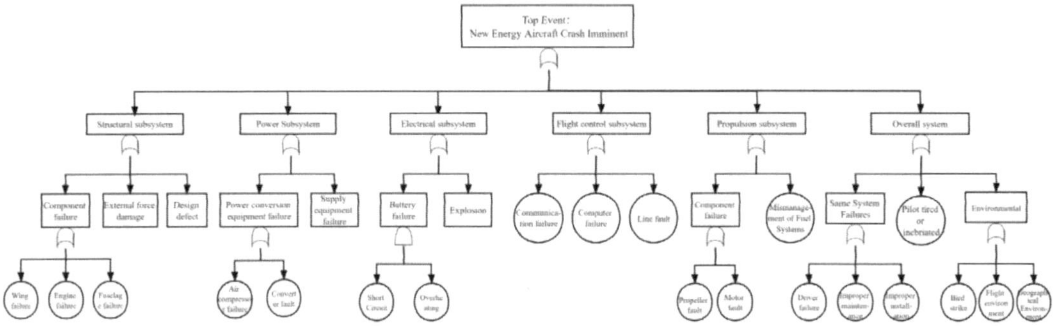

Figure 5. Crash Imminent Fault Tree

From the FMEA, it was concluded that most failure causes of components fall under four possible scenarios: Human or Operator Error, Manufacturing Defects, Material Defects, and Environmental Effects. All other possible failures come under Technical Failures of any component of the system,

making a total of five categories. Hence, individual component failures could be classified into any one of these five categories. It was also able to be seen that subsequent failures could be grouped under their respective subsystems. Not all types of failures apply to all subsystems, such as the Human Safety subsystem not depending on any Manufacturing Defects. After assessing each individual component failure, a separate group for Overall system failures was created.
Another fault tree was created to visualize and better understand if a top event was a crash imminent for the new energy aircraft (as figure 5).

5. Summary and expectation

Given that the new energy aircraft system is still under development, obtaining reliable data has proven to be challenging. To address this issue, we have divided the overall system into several subsystems and compared them with traditional aircraft in this study. Additionally, we conducted a comprehensive analysis of failure modes for both the entire system and its subsystems using methods such as FHA, FMEA, FTA etc., completing the safety analysis. Currently, ensuring the reliability and safety of new energy aircraft is of utmost importance, while simultaneously addressing technical challenges to minimize potential faults in accordance with secure development requirements.

Reference

[1] Chen Nongtian, Ning Weifeng, Chen Junxi, et al. 2022 The Progress on Airworthiness Safety of New Energy Aircraft *Journal of civil aviation flight university of China* **33** 5-9
[2] Li Yuan, Lei Tao, Wang Hongxiao, et al. 2023 Review on the development of electric aircraft power system standards *Aeronautic Standardization & Quality* **1** 1-5+49
[3] Frederic P. 1996 *Guideline and Methods for Conducting the Safety Assessment Process on Civil Airborne Systems and Equipment*
[4] Ji Yuhan, Sun Xiaosheng, Yu Xiao, Wu Jiaxi, et al. 2022 Development Prospect of New Energy Aviation Under Carbon Peaking and Carbon Neutrality Goals *Aeronautic Science & Technology* **33** 1-11
[5] Civil Aviation Administration of China. 2022 CCAR-23-R4
[6] Zheng Youshi, Sun Youchao, Wang Fengchan, Wang Lin 2010 Study on Application of Safety Analysis Methods in Airworthiness Compliance on Civil Airplane *Conf. on Aircraft Airworthiness and Air Traffic Management (Beijing)* **37** 157-163
[7] Xu Weicheng, Luo Qiufeng, Gao Yanhui, ZhangRui 2020 Light Airworthiness UAV System Safety Analysis Process *Environment Adaptability & Reliability* **38** 71-77
[8] Zhang Zhuoran, Yu Li, Li Jincai, et al. 2017 Key Technologies of Advanced Aircraft Electrical Machine Systems for Aviation Electrification *Journal of Nanjing University of Aeronautics & Astronautics* **49** 622-634
[9] Li Kaixing 2019 Summary of Research on Core Technology of Electric Aircraft *Aeronautic Science & Technology* **30** 8-17
[10] A. Kozakiewicz, T. Grzegorczyk 2021 Electric Aircraft Propulsion *Journal of KONBiN* **51** 49-66
[11] B. J. Brelje, J. R. R. A. Martins 2018 Electric, Hybrid, and Turboelectric Fixed-wing Aircraft: A review of concepts, models, and design approaches *Progress in Aerospace Sciences* **104** 1-19
[12] Hu Yaohun, Wang Yang, Jin Tao, Liu Jinsong, Jin Yong 2018 Application of Safety Assessment Method in General Aircraft System *Telecommunication Engineering* **58** 1034-1039
[13] Liu Luxuan, Wang Lei, Ouyang Jingpeng 2022 Hu Yaohun, Wang Yang, Jin Tao, Liu Jinsong, Jin Yong 2018 Analysis of Airworthiness Requirements for Multi-Electric Aircraft Environment Control Systems *Science & Technology Vision* **9** 20-23
[14] Huang Jun and Yang Fengtian 2016 Development and challenges of electric aircraft with new energies *Acta Aeronautical et Astronautical Sinical* **37** 57-68
[15] Yang Fengtian, Fan Zhenwei, Xiang Song, Liu Yuanqiang, Zhao Weiping 2021 Technical Innovation and Practice of Electric Aircraft in China *Acta Aeronautical et Astronautical Sinical* **42** 624619

Optimization and verification of mechanical design for small total reflection optics

Shuai Dong*, Yangbo Xu, Youwei Ma, Libin Li and Shengsheng Han

Shanghai Aerospace Control Technology Institute, China

* Corresponding author's email: dongshuai3791@163.com

Abstract. Mechanical performance assurance is one of the key concerns in the design of small total reflection optics. This article addresses the issue of insufficient strength margin in the design of a total reflection optics for a space photoelectric sensor. Firstly, structural design optimization research was conducted on the optics, and the effectiveness of structural improvement was verified through finite element analysis; Secondly, a concave analysis was conducted under mechanical test conditions to further reduce the risk; Afterwards, the accuracy of the simulation results was verified by comparing the simulation with the experiment through small-scale random vibration tests; Finally, sine vibration tests, random vibration tests, shock tests, acceleration tests were conducted on the entire machine, and optical tests were conducted before and after the tests. The experimental and testing results indicate that the entire machine has passed the mechanical test assessment at the appraisal level and maintained good optical performance. The total reflection optics optimized by mechanical design meets the requirements of engineering applications.

1. Introduction

Optical payloads are widely used in satellites such as surveying, reconnaissance, and environmental monitoring[1,2], often with significant weight and volume[3]. With the demand for lightweight satellites and the development of commercial small satellites, there is a trend towards miniaturization of payloads[4]. The commonly used total reflection optical system for optical loads has gradually been applied to space photoelectric sensors represented by search cameras and star trackers after miniaturization design[5]. The total reflection optical system has the advantages of compact structure, wide spectral band, and no color difference compared to the commonly used transmission optics in the past, which is conducive for improving the imaging quality of single unit products and has broad application prospects[6].

The mechanical design of a total reflection optics has a significant impact on the structural stability of a single unit. Shanliang Ke designed an ultra light stable main support structure for ultra light space cameras and optimized it. The results showed that this structure can ensure the overall stability of the entire machine[7]. Wei Li optimized the support structure of the primary and secondary lens of the space camera and verified its structural stability through dynamic environment experiments[8]. Quanfeng Guo optimized the structural design of the support for the main lens and focusing lens of the coaxial three optics camera, and the experimental results showed that the structural stability of this space camera was good, meeting the indicator requirements[9].

After the miniaturization design of the optics, changes in support form, component configuration, and other factors may pose risks to the mechanical properties. This article focuses on the problem of insufficient strength margin in the design of a total reflection optics structure for a space photoelectric sensor, and optimizes the structural design. The concave conditions of the mechanical test were determined through exploratory testing, and the simulation results showed that the strength margin of the optics was high. The effectiveness of the structural improvement was verified through comparative analysis between the experiment and simulation.

2. Analysis of the original design

2.1. Optics overview

The total reflection optics used in a space photoelectric sensor adopts one-axis four-optics structure as shown in Figure 1 (a), and its envelope size is Φ 142mm × 70mm, with a weight of 1.155kg. It has the characteristics of compact structure, light weight, and easy installation and adjustment. The reflector is supported by flexible joints, as shown in Figure 1 (b). Its main function is to eliminate the influence of internal stress caused by structural processing and installation factors, but at the same time, there is also a risk of reducing support stiffness, excessive response during vibration, and significant increase in stress at flexible joints. Therefore, comprehensive consideration is needed for design.

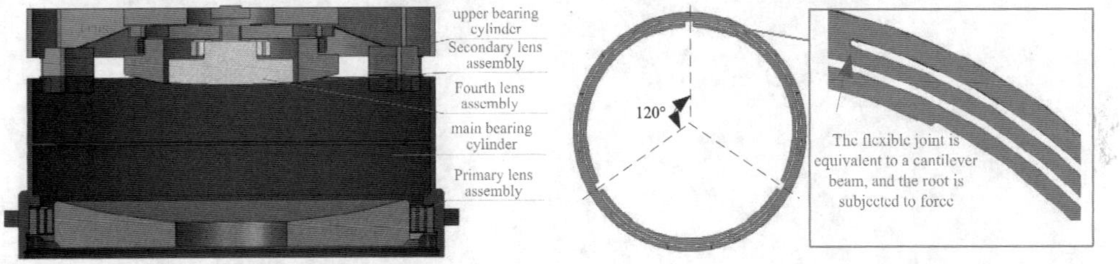

(a) Optics model (b) The flexible joint

Figure 1. Schematic diagram of optics model

2.2. Simulation analysis

To ensure a good mechanical interface for the installation of spatial photoelectric sensors, it is generally required that the first resonance frequency of the entire machine shall be higher than 100Hz. Conduct modal analysis of the entire machine using finite element analysis software. After analysis, the first-resonance frequency is 206.02Hz, with high stiffness. The random vibration frequency range is generally required to be within 2000Hz, and the resonance frequency of the optics part within this range is shown in Table 1. There are multiple resonance points within this range, so it is necessary to analyze the maximum stress of the optics in random vibration.

Table 1. Modal analysis results of optics

Order	**resonance frequency(Hz)**	**position**
1	606	Secondary lens assembly
2	832	Primary lens assembly
3	973	Upper bearing cylinder
4	1110	Primary lens assembly
5	1317	Upper bearing cylinder
6	1774	Secondary lens assembly

The random vibration test conditions for the entire machine appraisal level are shown in Table 2. Random vibration simulation analysis was conducted on the entire machine, and the stress cloud diagram of the component where the maximum stress of the optics in different directions is located is shown in Figure 2. In addition, the stress at the connection between the upper bearing cylinder and the main bearing cylinder, as well as the truss, is also large, as shown in Figures 3 and 4.

Journal of Physics: Conference Series **2633** (2023) 012007 doi:10.1088/1742-6596/2633/1/012007

Table 2. The level of the random vibration

Qualification Random vibration-X,Y,Z axes	
Freq.[Hz]	Level
20~100	+3dB/oct
100~600	$0.25g^2/Hz$
600~2000	-9 dB/oct
14.33gRMS	

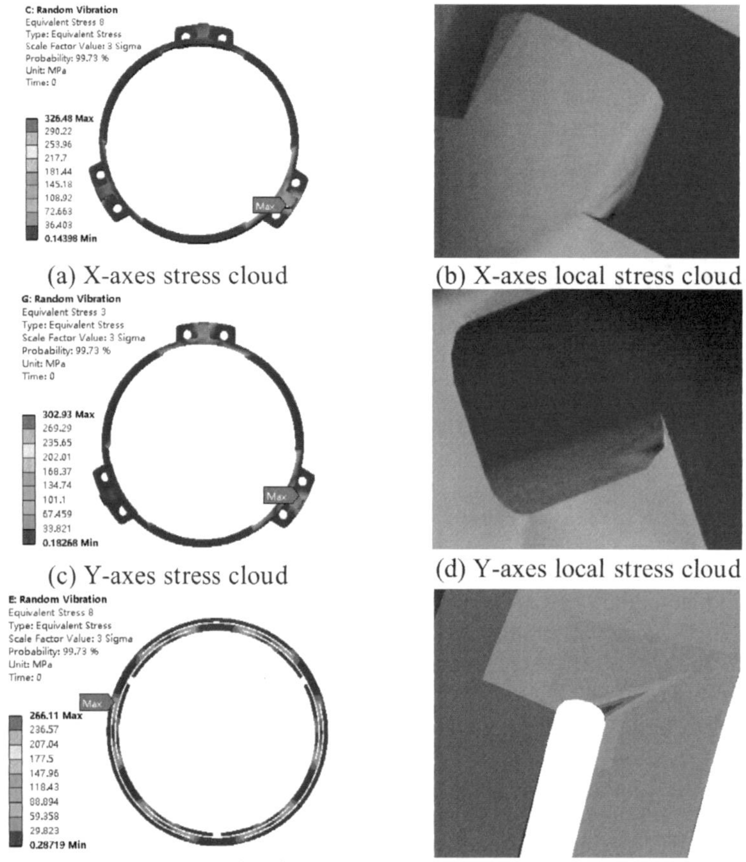

(a) X-axes stress cloud (b) X-axes local stress cloud

(c) Y-axes stress cloud (d) Y-axes local stress cloud

(e) Z-axes stress cloud (f) Z-axes local stress cloud

Figure 2. Random vibration analysis of the optics

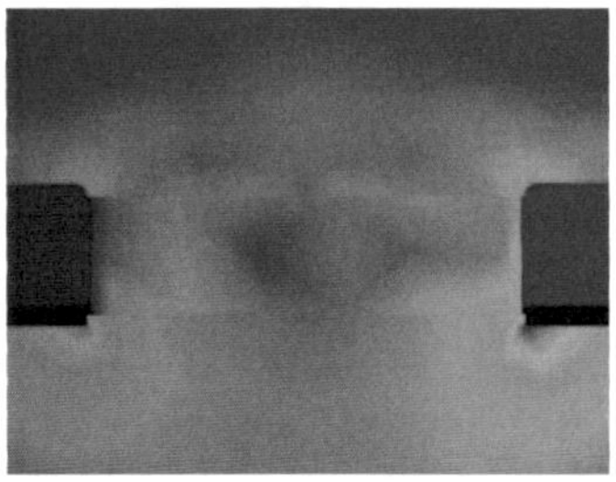

Figure 3. Local stress at the connection between the upper and main bearing cylinders

Figure 4. Local stress of the truss

According to the analysis results, the maximum stress in the X and Y directions of the random vibration reached 326.48MPa and 302.93MPa, which occurred at the secondary mirror frame, while the maximum stress in the Z direction was 266.11MPa, which occurred at the primary lens frame. Both the primary and secondary frame materials are made of invar steel. The yield stress of invar steel is 279MPa, and the allowable stress is 186MPa (with a safety factor of 1.5). The maximum stress obtained from simulation analysis has far exceeded the allowable stress, so the mechanical properties do not meet the requirements.

3. Design improvements

3.1. Structural improvement

Based on the aforementioned simulation analysis, it is known that:
(1) The maximum stress of random vibration occurs at the secondary mirror component, and the stress at the primary mirror component is also relatively high;
(2) The stress at the connection between the upper and the main bearing cylinder is relatively high.
(3) The stress at the root of the truss is relatively high.
Therefore, the following design improvements are conducted as Figure 5~Figure 8.

(a) Schematic diagram of the secondary lens before improvement

(b) Schematic diagram of the secondary lens after improvement

Figure 5. Improvement of the secondary lens

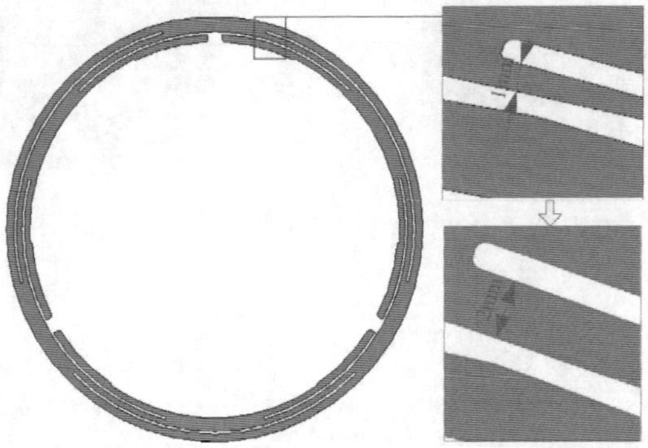

Figure 6. Improvement of the primary lens frame

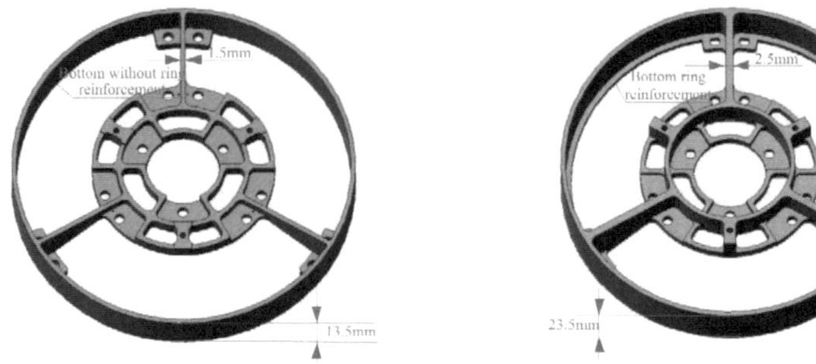

(a) Forward schematic diagram before improvement (b) Forward schematic diagram after improvement

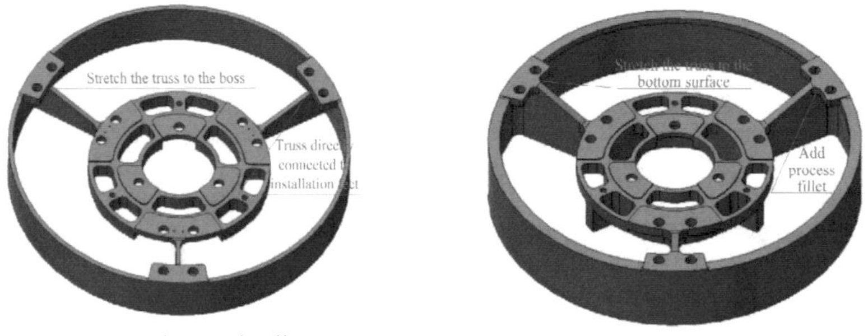

(c) Reverse schematic diagram before improvement (d) Reverse schematic diagram after improvement

Figure 7. Improvement of the upper bearing cylinder

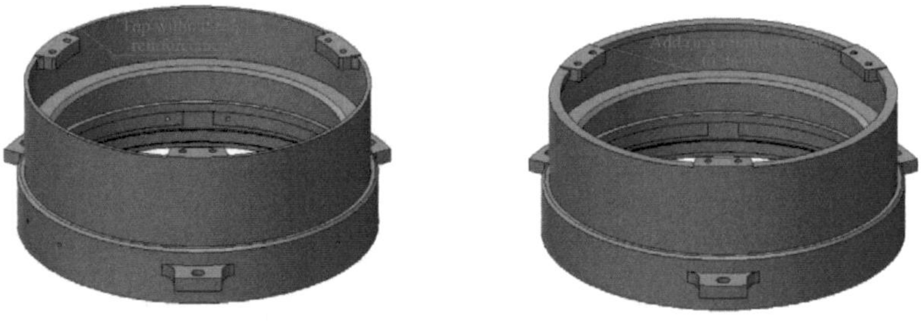

Figure 8. Improvement of the main bearing cylinder

3.2. Design verification

Qualification simulation analysis of the entire machine is conducted, and the optics simulation analysis results are shown in Figure 9. The comparison of the maximum stress of random vibration of the optics before and after structural improvement is shown in Table 3.

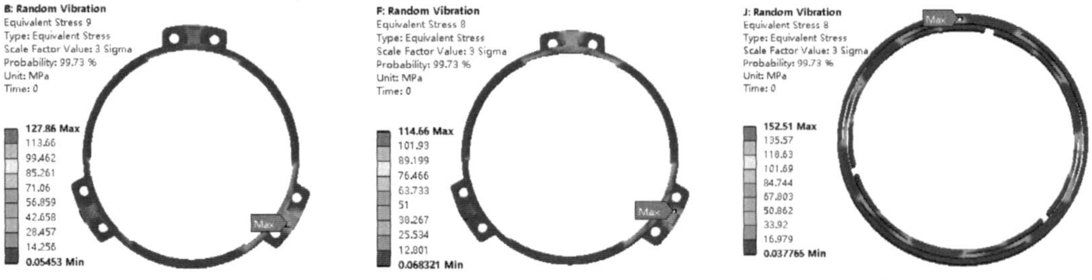

(a) X-axes stress cloud (b) Y-axis stress cloud (c) X-axis stress cloud

Figure 9. Stress cloud in random vibration of the optics

Table 3. Comparison of maximum stress before and after optics design optimization

direction	stress before optimization(MPa)	stress after optimization(MPa)	Variation
X	326.48	127.86	60.8%
Y	302.93	114.66	62.1%
Z	266.11	152.51	42.7%

The calculation formula for safety margin is as follows:

$$M.S. = S_a / S_e - 1 \qquad (1)$$

In the formula, M.S. is the safety margin, and the general metal requirement is greater than 0.15; Sa is the allowable stress, and Se is the stress generated by the qualification load. According to equation (1), the safety margin of random vibration of the optics is above 0.22. The safety margin of strength in the X, Y and Z directions meets the requirements. According to the table 3, the maximum stress decreased by 42.7%~62.1%, indicating that the structural improvement is effective. Due to possible deviations between simulation and testing, in order to further reduce risks, it is necessary to concave the mechanical test conditions to reduce the maximum stress at the optics component.

3.3. Concave analysis under mechanical test conditions

3.3.1. Determination of concave test conditions

In order to determine the concave frequency point, sine vibration simulation was conducted on the optics component, and the peak response points are shown in Table 4. Conduct concave processing near the response peak frequency point, and determine the concave test conditions as shown in Table 5, Table 6, and Table 7.

Table 4. Response peak frequency point

direction	X	Y	Z
resonance frequency(Hz)	562	521	626

Table 5. Random vibration conditions after concavity

Qualification Random vibration-X,Y,Z axes	
Freq.[Hz]	Level
20~100	+3dB/oct
100~460	$0.25g^2/Hz$
460~705	$0.07\ g^2/Hz$
706	$0.154\ g^2/Hz$
706~2000	-9 dB/oct
12.915gRMS	

3.3.2. Analysis of Random Vibration after Concave

Perform random vibration simulation based on the previously determined concave test conditions, and the analysis results are shown in Figure 10. The maximum stress comparison of random vibration of the front and rear concave lenses is shown in the table. According to equation (1), the safety margin is above 0.522, and the strength safety margin in the X, Y, and Z directions is relatively high. From the results in Table 6, it can be seen that the maximum stress after indentation has decreased by more than 19.9%, indicating that the indentation test conditions are effective.

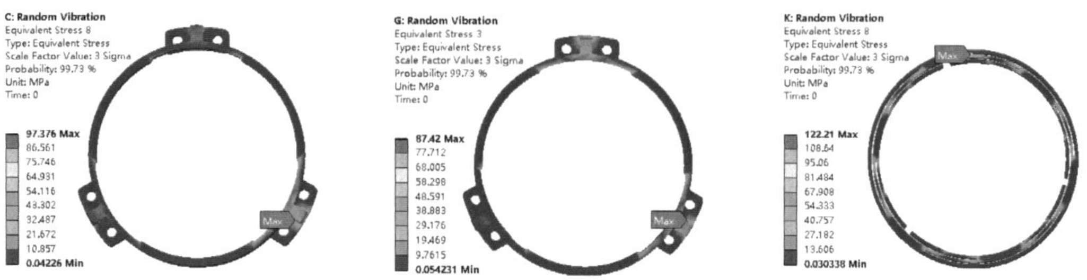

(a) X-axis stress cloud (b) Y-axis stress cloud (c) Z-axis stress cloud

Figure 10. Random vibration of the optics after concave

Table 6. Comparison of maximum stress of the optics before and after concave

direction	Maximum stress of the optics before concave(MPa)	Maximum stress of the optics after concave(MPa)	Variation
X	127.86	97.38	23.8%
Y	114.66	87.42	23.7%
Z	152.51	122.21	19.9%

4. Test verification

4.1. Small-scale test

In order to verify the accuracy of simulation results, it is necessary to compare and analyze their results with experimental results. Conduct small-scale random vibration tests on the lens components, and the position of the acceleration sensor is shown in Figure 11. Considering the operability of sticking strain measuring points during the experiment, the sticking position should be changed during the experiment, as shown in Figure 12, located at the root of the truss, with a total of four positions.

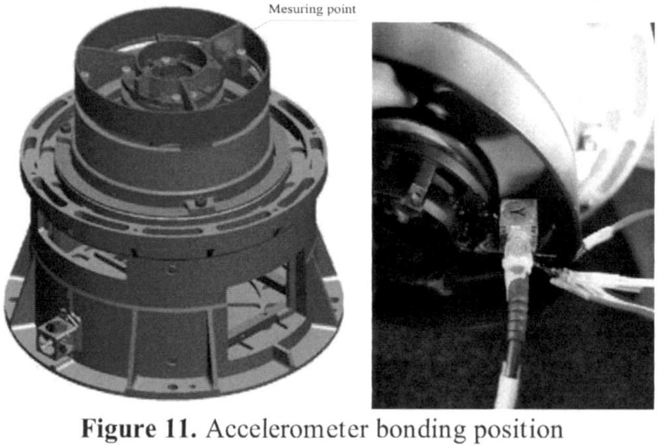

Figure 11. Accelerometer bonding position

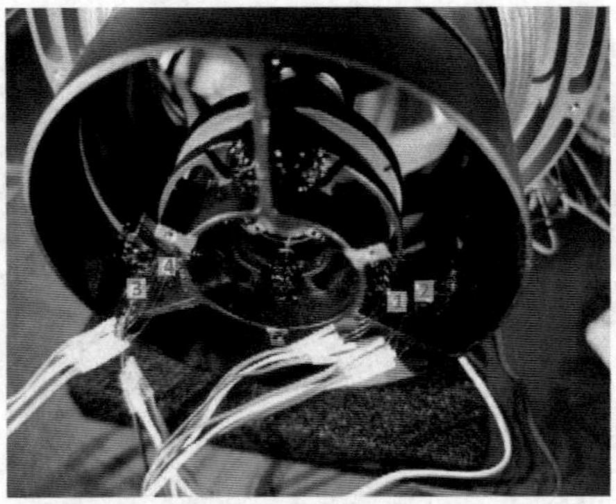

Figure 12. Strain rosette pasting position

Perform small-scale random vibration simulation analysis on the optics components, as shown in Figure 13. The comparison between the experimental measured stress and the simulation analysis results is shown in Table 7. The measured stress values near the root of the truss basically correspond to the simulation results, indicating that the simulation results are instructive.

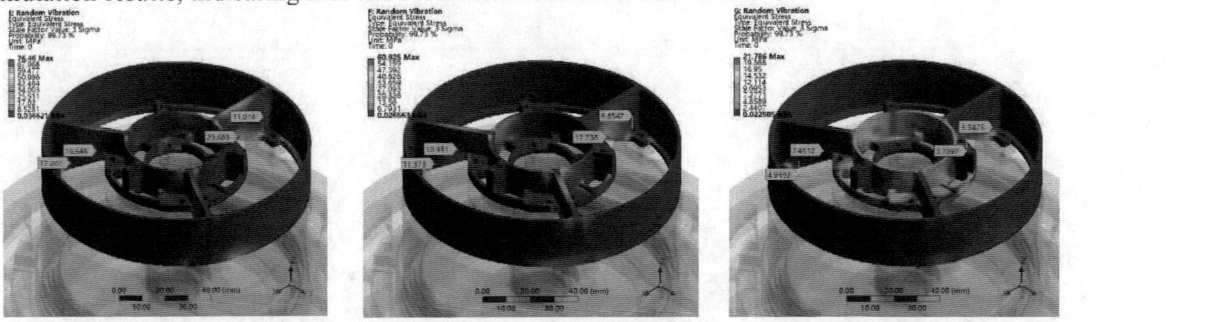

(a) X-axis stress cloud (b) Y-axis stress cloud (c) Z-axis stress cloud

Figure 13. Analysis of small-scale Random Vibration

Table 7. Comparison between experimental measured stress values and simulation analysis

position	X-axis test (MPa)	X-axis simulation (MPa)	X-axis error	Y-axis test (MPa)	Y-axis simulation (MPa)	Y-axis error	Z-axis test (MPa)	Z-axis simulation (MPa)	Z-axis error
point 1	18.66	16.55	11.31%	9.81	10.48	6.83%	8.09	7.45	7.91%
point 2	17.16	17.21	0.29%	11.99	11.37	5.17%	4.68	4.94	5.56%
point 3	11.75	11.07	5.79%	6.08	6.85	12.66%	3.36	3.35	0.30%
point 4	22.35	23.68	5.95%	17.61	17.74	0.74%	3.97	3.71	6.55%

4.2. Mechanical test verification

The entire machine was subjected to sine vibration tests, random vibration test, shock vibration test, and acceleration tests. The test site images are shown in Figure 14. Among them, the random vibration test conditions are concave, and the concave frequency points are further accurately obtained through sine vibration tests. The comparison between the concave frequency point test and simulation results is shown in Table 8. The deviation is minimal, which verifies the accuracy of the simulation results.

Figure 14. Qualification mechanical tests site

Table 8. Comparison of concave frequency points between simulation and experiment

direction	simulation results(Hz)	test results(Hz)	variation
X	562	549	2.4%
Y	521	512	1.7%
Z	626	606	3.3%

Conduct photo testing on the entire machine before and after the mechanical test, as shown in Figure 15 and Figure 16. The calibration accuracy test results before and after the mechanical test are shown in the table 9. From the figure, it can be seen that there is no significant change in the shape of the star point image after the mechanical test. From the table, it can be seen that there is no significant change in the optical calibration parameters and accuracy before and after the mechanical test, which meets the indicator requirements.

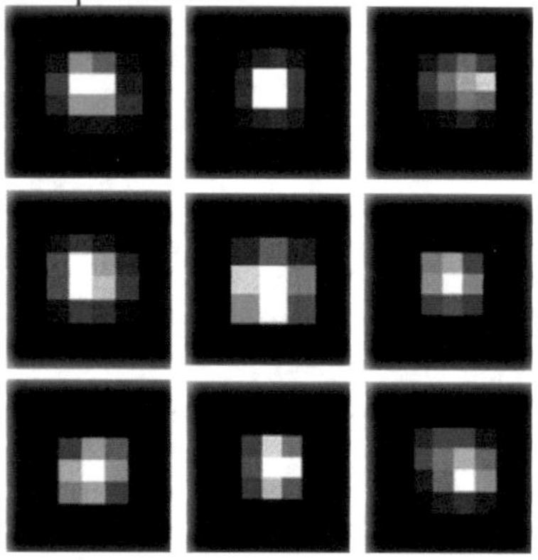

Figure 15. Morphology of star spots before experiment

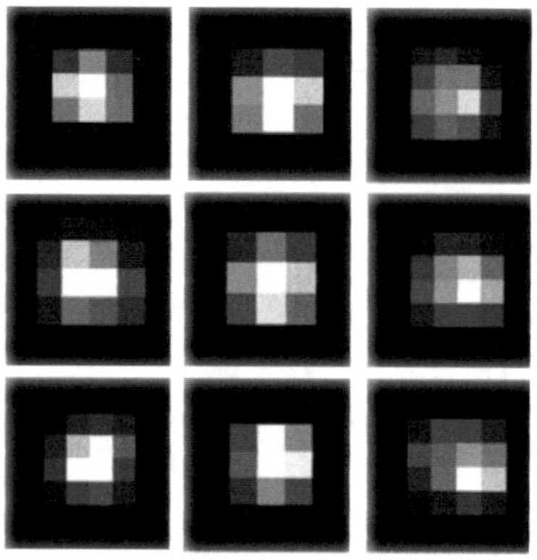

Figure 16. Morphology of star spots after experiment

Table 9. Comparison of calibration accuracy tests before and after mechanical tests

test time	Calibration parameters	accuracy(X,Y)
before mechanical tests	focal length:188.6190mm principal point:(X,Y):1013.8pix,1028.7pix	0.3", 0.3"
after mechanical tests	focal length:188.6148mm principal point:(X,Y):1012.7pix,1029.9pix	0.3", 0.3"

5. Conclusion

This article conducts mechanical design optimization research on the total reflection optical lens used in space photoelectric sensors. Firstly, the maximum stress of the original design state of the optics was obtained through mechanical simulation. Based on this analysis, it was found that the design strength margin of the secondary mirror component, primary mirror component, upper bearing cylinder, and main bearing cylinder was insufficient. As a result, the structure of the optics was improved, and the maximum stress of the improved optics decreased by more than 42.7%; Secondly, the concave conditions of the random vibration mechanics test were determined through sine vibration simulation. After the concave, the maximum stress of the optics component was reduced by more than 19.9%, and the safety margin was above 0.522. Subsequently, a comparative analysis was conducted between the experimental and simulation results, and the error was basically within 10%, indicating that the simulation results are instructive. Finally, mechanical tests were conducted on the entire machine, and the improved total reflection optics passed the qualification level mechanical test assessment, maintaining good optical performance and meeting application requirements.

References

[1] WU Han, JIN Lei. Multi-Scene High-Resolution Remote Sensing Image Fast Mosaic Technology[J]. Flight Control & Detection, 2021, 4(5): 62-70.

[2] ZHOU Yongjun, QIANG Li, LI Yuanxiang. Status and Development Trends of Foreign Space-Borne High-Resolution Imaging Remote Sensing Systems[J]. Flight Control & Detection, 2021, 4(5): 1-8.

[3] Douglas N, Angeli G, Claver C, et al. Overview of the LSST active optics system[J]. Proceedings of SPIE, 2014, 9150: 91500G.

[4] WU Yanxiong, WANG Liping. Optical system of star sensor with miniaturization and wide spectral band[J]. Journal of Applied Optics, 2021, 42(5): 782-789.

[5] WANG Yongxian, ZHU Junqing, ZHANG Ruiping. Design of Dexterous Four-Optical lens Star Sensor with Large Aperture[J]. Flight Control & Detection, 2022, 5(5): 6-9.

[6] REN Bing-Wen, HU Yan, JIN Guang, et al. Design of On-Axis Four-Optical lens Optical System for Earth Observation Small Satellite[C]. 2013 Symposium on Space Optics and Electromechanical Technology, Xi'an, 2013.

[7] KE Shanliang. Optimal design of the main support structure of ultra-light space camera[D]. Changchun: Changchun Institute of Optics，Fine Mechanics and Physics，Chinese Academy of Science，China，2018.

[8] LI Wei. Technology Studying On The Supporting Structure Between Primary Optical lens And Second Optical lens On The Space Camera.[D]. Changchun: Changchun Institute of Optics, Fine Mechanics and Physics, Chinese Academy of Science, China, 2018.

[9] GUO Quanfeng,. Research on Structure Stability of Three-Optical lens Cassegrain Space Camera[D]. Changchun: Changchun Institute of Optics, Fine Mechanics and Physics, Chinese Academy of Science, China, 2018.

Research on radial velocity and airspace position calculations of target in airborne active radar simulation

Wenguo Zhang[*, 1], Hanghua Gan[1], Xin Pan[1], Feng Tian[1], Li Pan[1], Peng Shao[1], Xiangle Jiang[1], Xiaobin Zou[2]

1. AVIC Chengdu Aircraft Industrial (Group) Co., Ltd., Chengdu 600092, China.
2. School of Materials Science and Engineering, Beijing Institute of Technology, Beijing 100081, China

*ORCID: 0000-0002-8259-7220 (Wenguo Zhang); Youngforzou@163.com (Xiaobin Zou).

Abstract. The detection of the target in the airborne active radar simulation is simplified to the calculations of the radial velocity, the maximum detecting range of the airborne active radar corresponding to the RCS, and the airspace position of the target. However, the coordinate transformations related to the calculations of the radial velocity and the airspace position are usually bewildering due to the different referenced coordinate systems. This paper introduces the radial velocity and airspace position calculations of the target in the airborne active radar simulation to illuminate the transformation process containing the specific rotation matrices, suggesting the unifying transformation to the north-up-east (NUE) rectangular coordinate system of the aircraft for both the radial velocity calculation and airspace position calculation of the target.

1. Introduction

Active radars, like the early warning radar and the fire-control radar, are used to detect the approximate or specific position of the moving target. There are two main active radar systems, the continuous-wave (CW) radar system, and the pulsed Doppler (PD) radar system. The CW radar can be divided into the simple CW radar, the frequency-modulated CW radar [1, 2], and the phase-modulated CW radar [3, 4]. The PD radars work on detecting objects based on measuring the Doppler shift of radar echoes through a coherent sequence of pulses and have been widely applied in applications requiring the detection of moving targets in a severely cluttered environment [5].

A tactical flight simulation usually consists of the state simulation, the flying environment simulation, the task system simulation, and the pilot controlling simulation to validate the availability of the tactical simulation system or the feasibility of the operational strategy. The airborne active radar simulation is a part of the task system simulation. It is an important simulation to test the usability of the airborne active radar system before the actual design and manufacturing.

The detection of the target is usually confirmed by the reflection wave of the active radar which is too difficult to implement the simulation due to the complex physical environment and signal processing [6, 7]. The reflection wave from the target will be detected by the airborne active radar when the radial velocity between the target and the aircraft is in the detecting range of the radial velocity of the airborne active radar [8]. The relative distance between the target and the aircraft should be below the maximum detecting range of the airborne active radar corresponding to the radar cross-section (RCS) of the target [9]. Furthermore, the prerequisite of the reflection wave is that the target is in the detectable airspace of the airborne active radar. Therefore, the detection of the target in the airborne active radar simulation should be simplified to the calculations of the radial velocity, the maximum detecting range of the airborne active radar corresponding to the target RCS, and the airspace position, just for supporting the tactical flight simulation system.

However, the coordinate transformations related to the radial velocity and the airspace position calculations are usually bewildering due to the different referenced coordinate systems. Thus, this paper introduces the radial velocity and airspace position calculations of the target in the airborne active radar simulation to illuminate the transformation process containing the specific rotation matrices, suggesting the unifying transformation to the north-up-east (NUE) rectangular coordinate system of the aircraft for both the radial velocity calculation and the airspace position calculation of the target.

2. Calculation data flow

The calculation data flow of our airborne active radar simulation is shown in Figure 1. The inputting parameters include the aircraft information, the target information list, and the antenna information. The aircraft information contains the aircraft's position, attitude, and north-up-east velocity. The target information list inputs each target's serial number, position, attitude, and speed scalar in every processing period of the calculation. The serial number is only used for marking the different targets with no contribution to any calculation.

The aircraft's position contains its longitude, latitude, and altitude under the geographic coordinate system. The NUE velocity is the vector composed of the components of its speed scalar in the three directions of the north-up-east system. The implication of the target's position is the same as the aircraft. The speed scalar refers to the moving direction of the target with no consideration of the angle of attack. If the head direction of the target is defined as the direction X then the component of the velocity vector in the direction X is equal to the speed scalar value, meaning the components in other directions are zeroes. The radial velocity between the aircraft and the target can be calculated by the above inputting information.

The airspace position is closely related to the orientation of the airborne radar since the airborne active radar is not all omnidirectional radar. The target will reflect the active radar wave only under the current scanning airspace of the antenna of the airborne active radar. The aircraft's position and attitude, the target's position, the antenna's attitude including the azimuth center and the pitching center, and its corresponding scanning range including the azimuth and the pitching range, are necessary for the airspace position calculation. The antenna radiation status notes the working mode of the left or the right radar of the aircraft, it does not participate in the airspace position calculation.

The calculation of the maximum detecting range of the airborne active radar corresponding to the target RCS is omitted in the data flow diagram because there is no coordinate transformation involved. The final calculations of each target would be confirmed for target detection. The detectable target would be outputted to the target traces list for the subsequent process of the data flow.

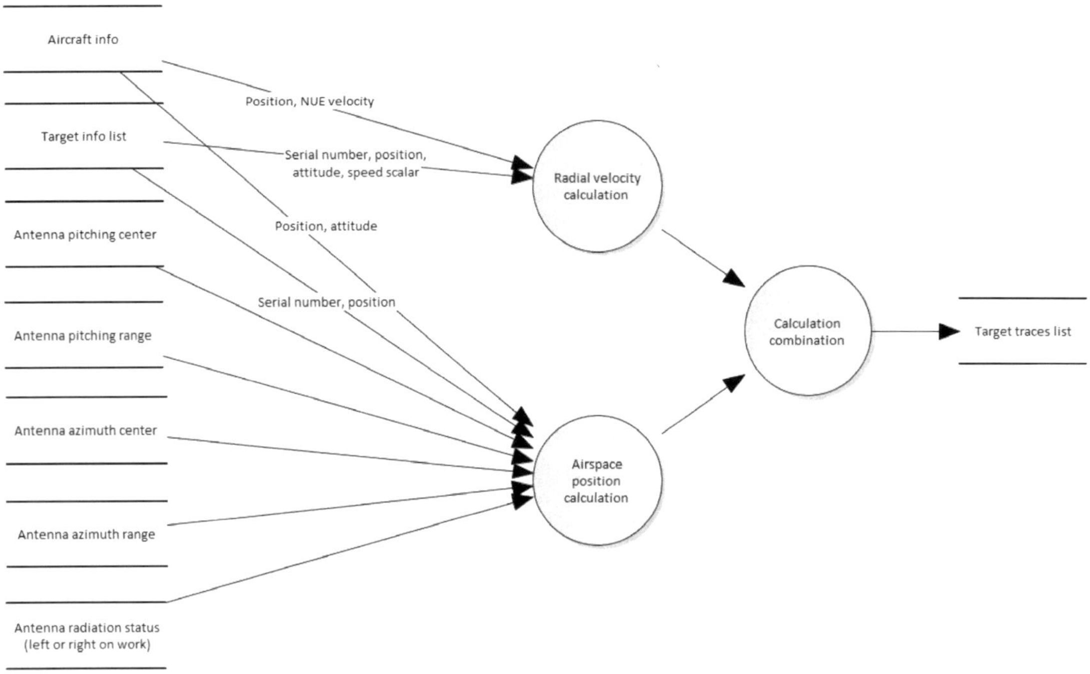

Figure 1. Calculation data flow diagram.

3. Coordinate transformation

3.1. The transformation from the geographic coordinate system to the WGS-84 system

$$\begin{cases} x = (N + c) \times \cos b \times \cos a \\ y = (N + c) \times \cos b \times \sin a \\ z = (0.99330562000987 \times N + c) \times \sin b \\ N = \frac{6378137}{(1 - 0.00669437999013 \times \sin^2 b)^{1/2}} \end{cases} \tag{1}$$

In equation set (1), a, b, and c represent the longitude, latitude, and altitude of the transformed object respectively. The calculated coordinate (x, y, z) is the coordinate transformed from the geographic coordinate system to the WGS-84 system.

3.2. The transformation from the WGS-84 system to the NUE system

The coordinate axis of the WGS-84 system can't be transformed to the NUE system of the aircraft by rotations of the longitude and latitude merely, as shown in Figure 2. There should be an axis aligning operation before the rotations of the longitude and latitude.

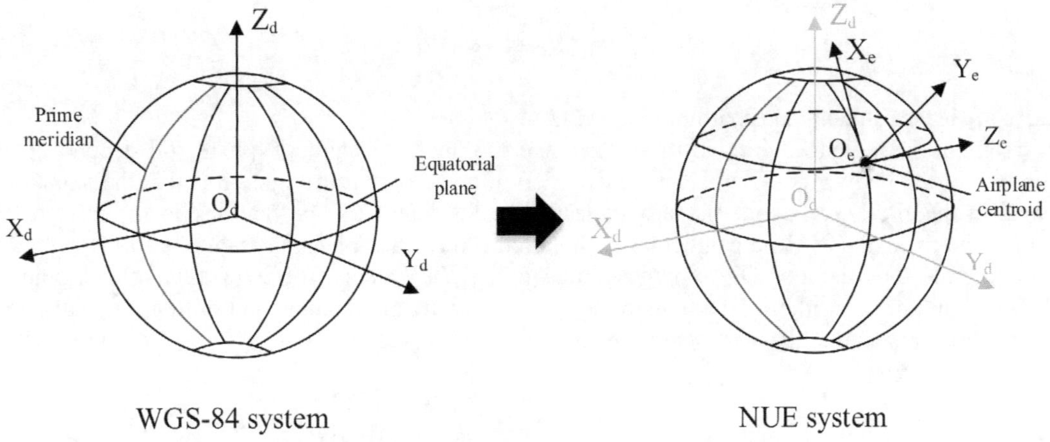

Figure 2. The transformation from the WGS-84 system to the NUE system.

$$A = \begin{bmatrix} 0 & 0 & 1 \\ 1 & 0 & 0 \\ 0 & 1 & 0 \end{bmatrix} \tag{2}$$

Matrix A is the coordinate axis aligning matrix from the WGS-84 system to the NUE system.

$$B = \begin{bmatrix} 1 & 0 & 0 \\ 0 & \cos\theta & \sin\theta \\ 0 & -\sin\theta & \cos\theta \end{bmatrix} \tag{3}$$

Matrix B is the longitude rotation matrix from the WGS-84 system to the NUE system where the θ is the longitude.

$$C = \begin{bmatrix} \cos\lambda & -\sin\lambda & 0 \\ \sin\lambda & \cos\lambda & 0 \\ 0 & 0 & 1 \end{bmatrix} \tag{4}$$

Matrix C is the latitude rotation matrix from the WGS-84 system to the NUE system where the λ is the latitude.

$$(x_N, y_N, z_N)^T = CBA(x, y, z)^T \tag{5}$$

The calculated coordinate (x_N, y_N, z_N) is the final coordinate transformed from the WGS-84 system to the NUE system.
The translation from the earth's core to the aircraft centroid is not considered yet, because this translation would be offset during the subsequent calculation of the relative position and velocity between the target and the aircraft.

3.3. The transformation from the NUE system to the body axis system

$$M_x = \begin{bmatrix} 1 & 0 & 0 \\ 0 & \cos\gamma & \sin\gamma \\ 0 & -\sin\gamma & \cos\gamma \end{bmatrix} \tag{6}$$

Matrix M_x is the rotation matrix around the axis X.

$$M_y = \begin{bmatrix} \cos\varphi & 0 & -\sin\varphi \\ 0 & 1 & 0 \\ \sin\varphi & 0 & \cos\varphi \end{bmatrix} \tag{7}$$

Matrix M_y is the rotation matrix around the axis Y.

$$M_z = \begin{bmatrix} \cos\xi & \sin\xi & 0 \\ -\sin\xi & \cos\xi & 0 \\ 0 & 0 & 1 \end{bmatrix} \tag{8}$$

Matrix M_z is the rotation matrix around the axis Z.
The relationships among the value of γ, φ, ξ, and the angle of azimuth, pitch, roll depend on the definition of the body axis system. The general definition as shown in Figure 3 is that the axis OX is located in the aircraft reference plane, parallel to the fuselage axis and pointing to the front of the aircraft, and the axis OY is perpendicular to the aircraft reference plane and pointing to the right side of the aircraft, then the axis OZ is perpendicular to the XOY plane in the reference plane, pointing below the aircraft. Considering the transforming order of yawing, pitching and rolling, the value of γ, φ, ξ would correspond to the angle of roll, pitch, and azimuth, respectively.

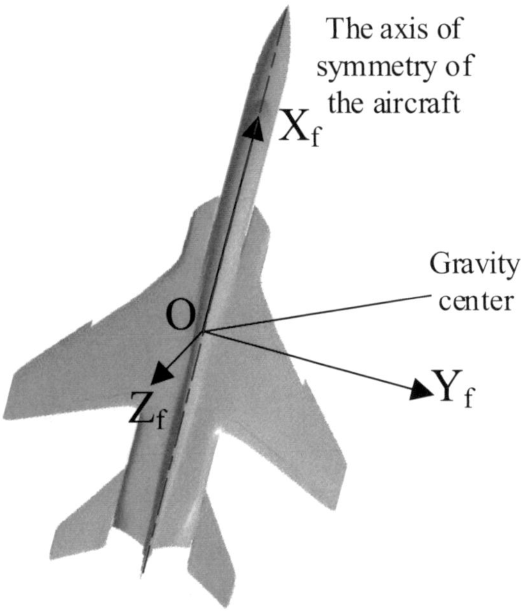

Figure 3. The general definition of the body axis system.

It should be noted that whether the NUE system can be transformed into the body axis system merely through the three rotation matrices depends on the definition of the body axis system. If the body axis system is defined as in Figure 3, there should be the additional aligned transformation matrix M_a from the NUE system to the body axis system of the aircraft as equation (9).

$$M_a = \begin{bmatrix} 1 & 0 & 0 \\ 0 & 0 & 1 \\ 0 & -1 & 0 \end{bmatrix} \tag{9}$$

Nevertheless, if the definition of the body axis system is correctly changed (meet the right-hand rule) by exchanging the axis OZ with the axis OY as Figure 4. The M_a should be the identity matrix as equation (10).

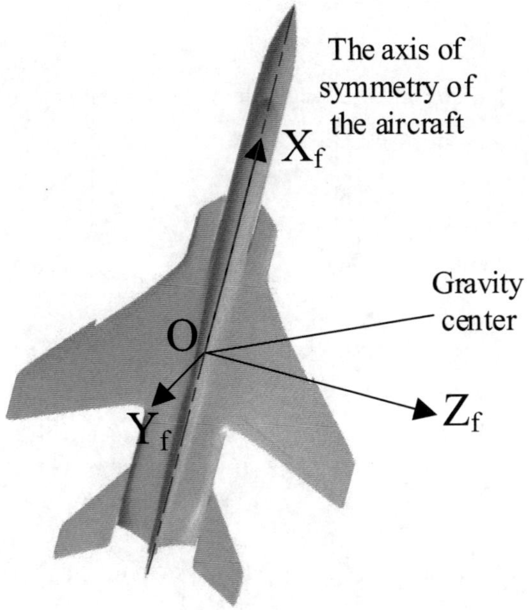

Figure 4. Another definition of the body axis system.

$$M_a = \begin{bmatrix} 1 & 0 & 0 \\ 0 & 1 & 0 \\ 0 & 0 & 1 \end{bmatrix} \tag{10}$$

The coordinate under the NUE system should be transformed by the additional transformation matrix M_a firstly before the subsequent axis rotations.

$$(x_b, y_b, z_b)^T = M_x M_y M_z M_a (x_N, y_N, z_N)^T \tag{11}$$

The coordinate (x_b, y_b, z_b) calculated by equation (11) is the final coordinate transformed from the NUE system to the body axis system.

4. Radial velocity and airspace position calculation

The body axis systems of the target, the aircraft, and the antenna in our airborne active radar simulation are both defined as Figure 3.

4.1. Radial velocity calculation

Firstly, the speed scalar of the target should be transformed to the velocity vector under the body axis system of the target. If the target's speed scalar is u then the velocity vector $\boldsymbol{v}$ under the body axis system of the target could be equation (12).

$$\boldsymbol{v} = (u, 0, 0)^T \tag{12}$$

The NUE velocity vector $\boldsymbol{v}'$ transformed from the body axis system to the NUE system of the target can be calculated by equation (13). The angle of γ, φ, ξ in corresponding M_x, M_y, M_z are the attitude angle of roll, pitch, azimuth of the target, respectively.

$$\boldsymbol{v}' = (M_x M_y M_z M_a)^{-1} \boldsymbol{v} \tag{13}$$

The transformation matrix M_t from the WGS-84 system to the NUE system of the target is equation (14). λ_t and θ_t represent the latitude and the longitude of the target respectively.

$$M_t = \begin{bmatrix} \cos\lambda_t & -\sin\lambda_t & 0 \\ \sin\lambda_t & \cos\lambda_t & 0 \\ 0 & 0 & 1 \end{bmatrix} \begin{bmatrix} 1 & 0 & 0 \\ 0 & \cos\theta_t & \sin\theta_t \\ 0 & -\sin\theta_t & \cos\theta_t \end{bmatrix} \begin{bmatrix} 0 & 0 & 1 \\ 1 & 0 & 0 \\ 0 & 1 & 0 \end{bmatrix} \tag{14}$$

The transformation matrix M_p from the WGS-84 system to the NUE system of the aircraft is equation (15). λ_p and θ_p represent the latitude and the longitude of the aircraft respectively.

$$M_p = \begin{bmatrix} cos\,\lambda_p & -sin\,\lambda_p & 0 \\ sin\,\lambda_p & cos\,\lambda_p & 0 \\ 0 & 0 & 1 \end{bmatrix} \begin{bmatrix} 1 & 0 & 0 \\ 0 & cos\,\theta_p & sin\,\theta_p \\ 0 & -sin\,\theta_p & cos\,\theta_p \end{bmatrix} \begin{bmatrix} 0 & 0 & 1 \\ 1 & 0 & 0 \\ 0 & 1 & 0 \end{bmatrix} \quad (15)$$

Then the NUE velocity vector $\boldsymbol{v}''$ of the target under the NUE system of the aircraft should be calculated by equation (16).

$$\boldsymbol{v}'' = (u_t, v_t, w_t)^T = M_p {M_t}^{-1} \boldsymbol{v}' \quad (16)$$

The target's position coordinate (x_t, y_t, z_t) and the aircraft's position coordinate (x_p, y_p, z_p) under the WGS-84 system can be calculated by the equation set (1) respectively. Then the relative position vector of the target under the WGS-84 system of the aircraft $\boldsymbol{p}$ can be obtained by equation (17).

$$\boldsymbol{p} = \left(x_t - x_p, y_t - y_p, z_t - z_p\right)^T \quad (17)$$

The relative position vector of the target under the NUE system of the aircraft $\boldsymbol{p_1}$ can be calculated by equation (18).

$$\boldsymbol{p_1} = M_p \boldsymbol{p} \quad (18)$$

If the NUE velocity vector of the aircraft $\boldsymbol{\mu} = \left(u_p, v_p, w_p\right)^T$, then the velocity vector difference $\boldsymbol{v_d}$ can be calculated by equation (19).

$$\boldsymbol{v_d} = \boldsymbol{v}'' - \boldsymbol{\mu} = \left(u_t - u_p, v_t - v_p, w_t - w_p\right)^T \quad (19)$$

Finally, the radial velocity scalar v_r can be calculated by equation (20) where the α is the intersection angle between $\boldsymbol{v_d}$ and $\boldsymbol{p_1}$ which is also called the aspect angle. Howbeit, it's not necessary to calculate α.

$$v_r = |\overrightarrow{v_d}|\,cos\,\alpha = \frac{\overrightarrow{v_d} \cdot \overrightarrow{p_1}}{|\overrightarrow{p_1}|} \quad (20)$$

The target is approaching the aircraft if the value of v_r is negative whereas it's leaving the aircraft if the value of v_r is positive. The target will be defined as radial velocity detectable if the calculated v_r is in the detectable range of the radial velocity in our simulation system.

4.2. Airspace position calculation

The angle of the aircraft's attitude should be defined before calculating the airspace position of the target. In the vertical view, the clockwise angle of the azimuth is defined as positive whereas the counterclockwise angle is defined as negative with the range of $-\pi$ to π. In the right-side view, the counterclockwise angle of the pitch is defined as positive whereas the clockwise angle is defined as negative with the range of $-\pi/2$ to $\pi/2$. In the head front view, the counterclockwise angle of the roll is defined as positive whereas the clockwise angle is defined as negative with the range of $-\pi$ to π. The definition of the aircraft attitude angle is presented in Figure 5.

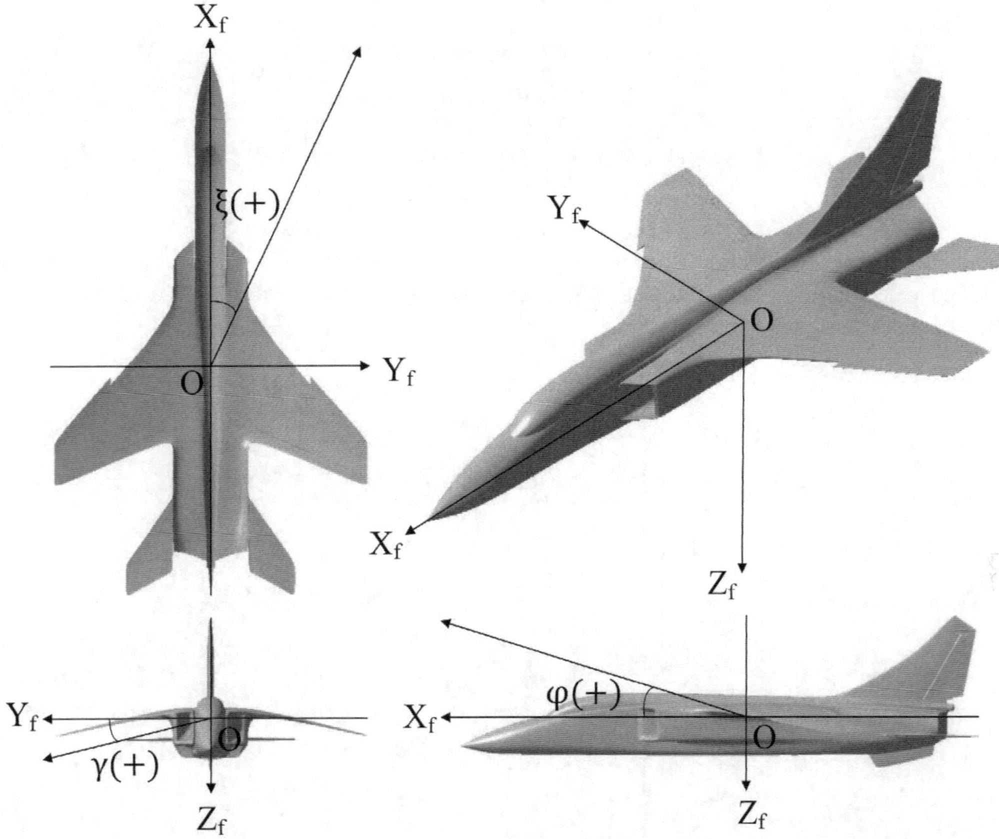

Figure 5. The definition of the attitude angle from the three views of the aircraft.

The target coordinate under the body axis system of the aircraft $\boldsymbol{p_2}$ can be calculated by equation set (21) where the γ_p, φ_p, ξ_p correspond to the angle of roll, pitch, and azimuth of the aircraft, respectively.

$$\begin{cases} \boldsymbol{p_2} = M_x{}' M_y{}' M_z{}' M_a \boldsymbol{p_1} \\ M_x{}' = \begin{bmatrix} 1 & 0 & 0 \\ 0 & \cos\gamma_p & \sin\gamma_p \\ 0 & -\sin\gamma_p & \cos\gamma_p \end{bmatrix} \\ M_y{}' = \begin{bmatrix} \cos\varphi_p & 0 & -\sin\varphi_p \\ 0 & 1 & 0 \\ \sin\varphi_p & 0 & \cos\varphi_p \end{bmatrix} \\ M_z{}' = \begin{bmatrix} \cos\xi_p & \sin\xi_p & 0 \\ -\sin\xi_p & \cos\xi_p & 0 \\ 0 & 0 & 1 \end{bmatrix} \\ M_a = \begin{bmatrix} 1 & 0 & 0 \\ 0 & 0 & 1 \\ 0 & -1 & 0 \end{bmatrix} \end{cases} \tag{21}$$

The target coordinate under the body axis system of the antenna $\boldsymbol{p_3}$ can be calculated by equation set (22) where the φ_a, ξ_a correspond to the angle of pitch and azimuth of the antenna respectively with no necessity to consider the rolling angle and the tilt angle during the installation of the antenna.

$$\begin{cases} \boldsymbol{p_3} = M_y{}''M_z{}''\boldsymbol{p_2} \\ M_y{}'' = \begin{bmatrix} \cos\varphi_a & 0 & -\sin\varphi_a \\ 0 & 1 & 0 \\ \sin\varphi_a & 0 & \cos\varphi_a \end{bmatrix} \\ M_z{}'' = \begin{bmatrix} \cos\xi_a & \sin\xi_a & 0 \\ -\sin\xi_a & \cos\xi_a & 0 \\ 0 & 0 & 1 \end{bmatrix} \end{cases} \tag{22}$$

Then the angle of azimuth and pitch of the target under the body axis system of the antenna should be calculated after obtaining the target coordinate $\boldsymbol{p_3}(x_a, y_a, z_a)$. The azimuth angle υ and the pitch angle ψ would be calculated by equation set (23) and equation set (24) respectively according to the definition of the aircraft attitude angle.

$$\upsilon = \begin{cases} \begin{cases} \tan^{-1}\left(\frac{y_a}{x_a}\right), x_a > 0 \\ \pi + \tan^{-1}\left(\frac{y_a}{x_a}\right), x_a < 0 \text{ and } y_a > 0 \\ -\pi + \tan^{-1}\left(\frac{y_a}{x_a}\right), x_a < 0 \text{ and } y_a < 0 \\ \pi, x_a < 0 \text{ and } y_a = 0 \end{cases} \\ \begin{cases} \frac{\pi}{2}, x_a = 0 \text{ and } y_a > 0 \\ -\frac{\pi}{2}, x_a = 0 \text{ and } y_a < 0 \\ 0, x_a = 0 \text{ and } y_a = 0 \end{cases} \end{cases} \tag{23}$$

$$\psi = \begin{cases} -\tan^{-1}\left(\frac{z_a}{\sqrt{x_a{}^2 + y_a{}^2}}\right), x_a{}^2 + y_a{}^2 \neq 0 \\ \begin{cases} -\frac{\pi}{2}, x_a{}^2 + y_a{}^2 = 0 \text{ and } z_a > 0 \\ \frac{\pi}{2}, x_a{}^2 + y_a{}^2 = 0 \text{ and } z_a < 0 \\ 0, x_a{}^2 + y_a{}^2 = 0 \text{ and } z_a = 0 \end{cases} \end{cases} \tag{24}$$

The target will be defined as airspace detectable if the calculated υ and ψ are in the detectable range of the azimuth and pitch of the antenna respectively in our simulation system.

5. Conclusion

The detection of the target in the airborne active radar simulation is simplified to the calculations of the radial velocity, the maximum detecting range of the airborne active radar corresponding to the RCS, and the airspace position of the target in this paper. The coordinate transformation related to the calculations of the radial velocity and the airspace position of the target is proposed. It suggests that the transformation should be unified to coordinate under the NUE system of the aircraft for both the radial velocity calculation and the airspace position calculation of the target. There is an additional aligned transformation matrix from the NUE system to the body axis system of the aircraft during the airspace position calculation, and the specific additional matrix depends on the definition of the body axis system of the aircraft.

The proposed coordinate transformation method can be used not only in the active radar simulation but also in the actual system containing the real-time calculation of the relative position of the target, for instance, the flight control in missile guidance, the target tracking in the Track-While-Scan (TWS) radar [10], etc.

Acknowledgments

This paper is dedicated to my beloved daughter Yunyi Zhang for her incredible birth which lights up my life.

References

[1] Ylaya V J V, Gerasta O J L, MacAsero J M S, Pongcol D P, Pandian N M and Vicerra R R P 2020 Linear Frequency Modulated Continuous Wave LFM-CW Short-Range Radar for Detecting Subsurface Water Content with Deep Learning. In: *12th IEEE International Conference on Humanoid, Nanotechnology, Information Technology, Communication and Control, Environment, and Management, HNICEM 2020, December 3, 2020 - December 7, 2020,* (Manila, Philippines: Institute of Electrical and Electronics Engineers Inc.)

[2] Muzhikyan A H 2011 Localization of Targets in CW-LFM Radars with Combined Modulating Signal *Journal of Contemporary Physics-Armenian Academy of Sciences* **46** 177-9

[3] Giannini V, Guermandi D, Shi Q X, Medra A, Van Thillo W, Bourdoux A and Wambacq P 2014 A 79 GHz Phase-Modulated 4 GHz-BW CW Radar Transmitter in 28 nm CMOS *Ieee Journal of Solid-State Circuits* **49** 2925-37

[4] An S, He Z S, Li J, An J and Zirath H 2020 Micrometer Accuracy Phase Modulated Radar for Distance Measurement and Monitoring *IEEE Sens. J.* **20** 2919-27

[5] Sun Y H, Fan H Y, Ren L X, Mao E and Long T 2022 Folded Clutter Suppression for Pulse-Doppler Radar Based on Pulse-Agile Waveforms *Ieee Transactions on Signal Processing* **70** 3774-88

[6] Wu Q and Zhang W 2014 Modeling and simulation of airborne radar clutter in a littoral complex environment. In: *2014 IEEE International Conference on Communication Problem-Solving, ICCP 2014, December 5, 2014 - December 7, 2014,* (Beijing, China: Institute of Electrical and Electronics Engineers Inc.) pp 496-9

[7] Shen F, Li L, Xu X, Wu R, Liu C, Zhou H and Guo Z 2017 Verification and analysis of electromagnetic environment based on Simulink. In: *2017 IEEE Region 10 Conference, TENCON 2017, November 5, 2017 - November 8, 2017,* (Penang, Malaysia: Institute of Electrical and Electronics Engineers Inc.) pp 1054-7

[8] Yu Y and Wu L 2013 A new method of velocity measurement based on CRT algorithm in dual-frequency pulse Doppler radar. In: *National Doctoral Academic Forum on Information and Communications Technology 2013, August 21, 2013 - August 23, 2013,* (Beijing, China: Institution of Engineering and Technology)

[9] DiFranco J and Rubin W 1980 *Radar Detection*

[10] Kazem A and Salut G 2008 Track-while-scan radar using deterministic Particle filtering. In: *2008 3rd International Conference on Information and Communication Technologies: From Theory to Applications, ICTTA, April 7, 2008 - April 11, 2008,* (Damascus, Syria: IEEE Computer Society)

Effect of oxygen rich environment on detonation characteristics of pulse detonation engine

M F Yang[1,*], S B Qi[1], H B Bai[1], S Y Gao[1], J Song[1], B W Hu[1] and Z Xu[1]

[1] Xi'an Modern Control Technology research Institute, Xi'an, Shanxi Province, Code 710000, China

* Corresponding author's email: njust_ymf@163.com

Abstract. The detonation tests of propane/air pulse detonation engine at normal temperature were carried out in this paper. The effects of chemical ratios on the detonation performance of propane/air pulse detonation engine in oxygen-rich environment were studied. The propagation modes and characteristics of detonation waves and the detonation characteristics of the engine under different conditions were analyzed. Detonation tests were conducted on propane/air mixtures with oxygen mass fraction of 20%, 25%, 30% and 35% respectively. The results showed that: In the detonation experiment of the propane/air mixture with oxygen content of 20%, 25% and 30%, the flame wave propagation velocity in the detonation tube is low, which fails to form a stable detonation wave. The detonation effect of the fuel mixture with oxygen content of 20% is the worst, and the propagation velocity of flame wave is 384.00 m/s. In the experiment of propane/air mixture with oxygen content of 35%, the detonation velocity and pressure obtained by pressure signal analysis are both greater than the theoretical CJ detonation velocity, and the detonation state is successfully reached.

1. Introduction

Pulse detonation engine is a new type of engine that generates high temperature and high pressure gas through pulse detonation wave to obtain thrust. It has the advantages of high thermal cycle efficiency, simple structure and so on. It has broad application prospects in aerospace, weapons and civil fields.

Propane is a typical fuel with advantages such as high activity and easy storage[1]. Its detonation characteristics are similar to aviation gasoline, and it has been taken as a research object by many scholars in the field of detonation combustion[2-4]. Ting J M[5] respectively studied the single detonation properties of C_3H_8/O_2 and C_3H_8/air mixtures, and used the detonation waves generated by C_3H_8/O_2 to ignite C_3H_8/air mixtures. The results show that the mixture of C_3H_8/O_2 generates detonation wave at a very short distance and can make the mixture of C_3H_8/air detonate. The distance between dedetonation and detonation (DDT) increased with the increase of pipe diameter. Li C M[6] explored the influence of pipe diameter conditions and nitrogen content on detonation characteristics of C_3H_8/O_2. The study showed that DDT distance was shortened with the decrease of pipe diameter for the chemically appropriate ratio and slightly rich C_3H_8/O_2 mixture. For the lean oil mixture, DDT distance does not decrease in the diameter of the tubule. Yu H[7] simulated the detonation process of hydrogen and air mixture from a one-dimensional enclosed space, and the experiment showed that the initial temperature, pressure and enclosed space size all had an impact on the generation of detonation waves, and the back and forth reflection of pressure waves could further shorten the DDT process. Lu F K[8] directly used propane as fuel to conduct a multi-cycle detonation test in a pulsed detonation engine.

In summary, many scholars have studied the single detonation and multi-cycle detonation of propane, which shows that propane has a great application prospect in detonation combustion. However, current researches on the characteristics of DDT under different initial parameters mainly focus on hydrogen and ethylene, and there is still a lot of work to be done on the initiation characteristics of propane and DDT. On this basis, we will conduct a study on the detonation characteristics of propane/air pulse detonation engine at normal temperature, carry out the pulse detonation experiment in different oxygen rich environment under the chemical proper ratio, and analyze the detonation wave propagation mode, propagation characteristics and the detonation characteristics of the engine.

2. Test system and test

The propane/oxygen-rich air detonation test system is shown in Figure 1, which mainly includes detonation tube, gas supply system, pressure signal data acquisition system, ignition control system and high-speed camera. The gas supply system uses standard propane/air/oxygen cylinders to supply gas to the test system. The diameter of the detonation tube is 30mm, the length of the tube is 600mm, and the spoiler with 150mm length is installed inside the tube, which is composed of 7 spoilers. The blocking ratio is 0.25, and the tail of the tube is sealed with PET film. The pressure signal data acquisition system uses PCB sensors, and the spacing between pairs is 90mm. The ignition control system uses high-energy igniter discharge, and the discharge energy is 1J. The detonation test system was carried out at normal temperature, and the gas temperature in the detonation tube was 302K and the pressure in the detonation tube was 0.1013MPa. Then, the hot state test of the pulse detonation engine was carried out in different oxygen-rich environments under the chemical proper ratio of propane/oxygen-rich air, and the oxygen mass fraction of the fuel in different oxygen-rich environments was 20%, 25%, 30% and 35%, respectively.

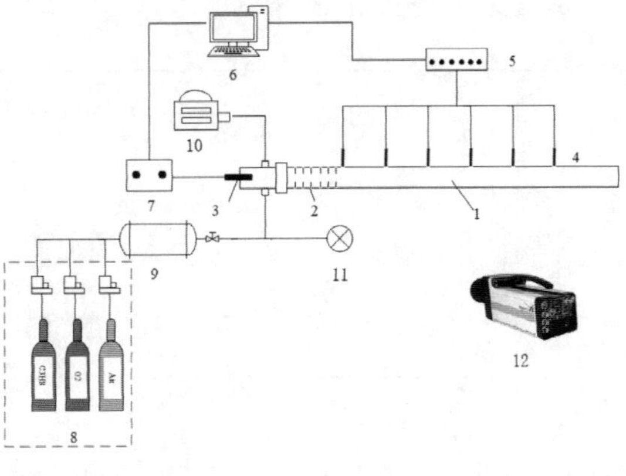

1. Detonation tube 2. Spoiler 3. Ignition head 4. Pressure sensor 5. Signal conditioning instrument 6. Signal acquisition system and ignition control system 7. High energy igniter 8. Air supply system 9 mixing chamber 10 vacuum pump 11 pressure gauge 12 high speed camera

Figure 1. Schematic diagram of propane/air detonation test system.

Propane, oxygen and air are filled into the premix chamber through the mass flow controller in accordance with the prescribed flow rate respectively. After the mixed gas is evenly mixed, fuel is filled into the head of the detonation tube, wherein the ratio of fuel and oxidizer complies with the appropriate chemical ratio of the complete combustion of propane. In the process of the experiment, the detonation tube was pumped by vacuum pump first, stopped when the pressure number was -0.098 MPa. According to the ratio of experimental conditions, the mass flow controller of propane, oxygen and air was turned on successively, and the mixed gas was fully mixed in the mixing chamber and charged into the detonation tube through the intake section of the detonation tube. Close the mass flow controller until the pressure indication number is 0 MPa, and then close the mixing chamber outlet valve. At this moment, the process of filling gas into the detonation tube is completed. A 5Hz ignition signal is sent out by a PC, and the mixed gas inside the detonation tube is ignited by a high-energy igniter. When the ignition head releases an electric spark, the mixed gas inside the detonation tube is ignited to form an initial flame wave, and the flame wave and leading shock wave rapidly propagate downstream to accelerate the chemical reaction of the mixed gas at the back end of the detonation tube, and DDT process is quickly realized under the action of the baffle. It's a transition from combustion to detonation. When the detonation wave propagates in the detonation tube, the PCB high-frequency dynamic sensor generates signals under the action of the dynamic pressure in the tube, which are fed back to the PC through the PCB signal conditioning instrument and high-speed data acquisition system. The D-10 channel signal acquisition and analysis system software is written by LabView software. During the experiment, the frequency of PCB pressure signal sensor data was 1000K Samples/s. The high-speed camera is placed 2 m in front of the detonation tube, and its height is consistent with that of the detonation tube. The frame number of the high-speed camera is set to 20000, and the exposure time depends on the experimental conditions.

3. Test results and analysist

In order to explore the influence of oxygen-rich environment on the detonation performance of propane/air pulse detonation engine, this paper carried out experimental studies on propane in air and oxygen-rich environment, and calculated the CJ detonation velocity vCJ and theoretical detonation pressure PCJ of propane in air and oxygen-rich environment combined with CEA software. The oxygen mass fraction of fuel in group A1, A2, A3 and A4 mixture is 20%, 25%, 30% and 35% respectively. As shown in Table 1.

Table 1. Theoretical values of detonation performance parameters of mixed gas.

number	Oxygen Volume fraction %	v_{CJ} /(m • s^{-1})	P_{CJ} /MPa
A1	20	1496	1.8
A2	25	1542	1.9
A3	30	1612	2.1
A4	35	1750	2.35

The four groups of gas mixtures in Table 1 are all proportioned in a certain proportion, and all conform to the appropriate chemical ratios of propane/hydrocarbon fuel in air and oxidizer in air. Under normal temperature and pressure, the above four groups of mixed gases were filled into the detonation tube for detonation experiment, and the experiment was repeated for 3 times to ensure the accuracy of the experimental results. In the experiment, the detonation tube was pumped to -0.098MPa, and the propane/air/oxygen mass flow controller was opened at the same time to fill gas into the gas mixing chamber, then the valve of the gas mixing chamber was opened to fill gas into the detonation tube, and the air inlet valve was closed when the pressure representation number was 0 MPa. The mixture ratio was in accordance with the volume fraction ratio in the above table. To ensure the accuracy of chemical equivalent ratio; Through the ignition control system, the ignition signal is sent to the high-energy igniter to ignite the mixed gas, and the flame wave accelerates to form a stable detonation wave propagating downstream under the action of the spoiler. The wave front quickly sweeps the unburned mixed gas, improves the gas temperature and pressure in the detonation tube, accelerates the chemical reaction of the mixed gas, and then forms a reflected shock wave propagating to the tube out of the tube port. A stable detonation is achieved. By analyzing the pressure signal of the PCB dynamic pressure sensor at the end face of the detonation tube, the signal is processed and analyzed by the high-speed data acquisition system, and the propagation velocity of flame wave in the tube and the change of pressure in the tube are obtained.

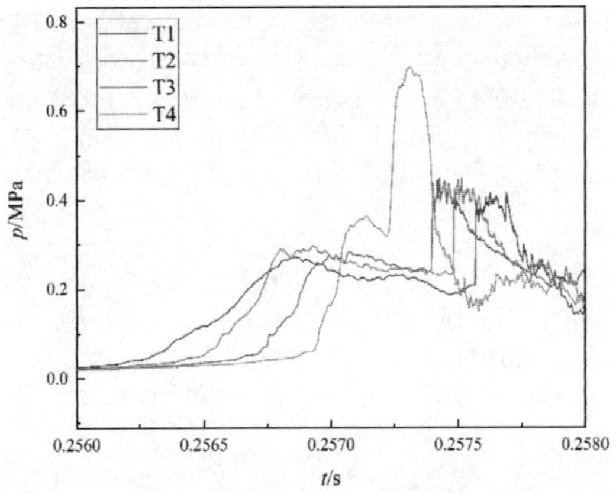

(a) A1 Group gas mixture

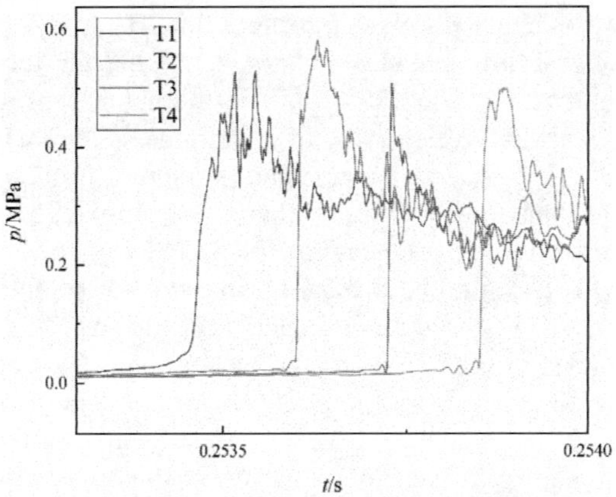

(b) A2 Group gas mixture

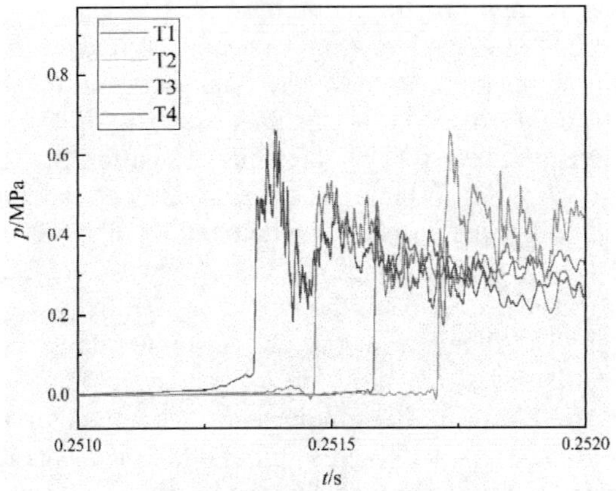

(c) A3 Group gas mixture

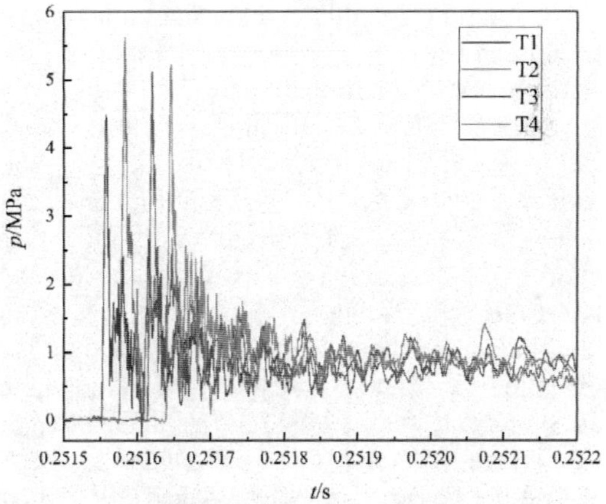

(d) A4 Group gas mixture

Figure 2. Pressure signal curves of mixed gas detonation tubes for each group.

Figure. 2 shows the pressure signal curve measured by the PCB dynamic pressure sensor after the ignition of propane/air/oxygen mixed gas with different ratios in the detonation tube under the above experimental conditions. In the figure, the four sensors (T1, T2, T3 and T4) are evenly distributed on the tube wall, and the sensor spacing (d) is 9 cm. Four pressure sensors record the time when the flame wave reaches the sensor respectively, and the flame wave velocity v and pressure p can be obtained by analyzing the interval time Δt-1 and the peak pressure between the sensor signals, in which the flame wave velocity formula $v=d\cdot\Delta t^{-1}$. Figure 2 (a), (b), (c) and (d) respectively show the pressure signal sensor curves in the detonation tube after ignition with four groups of mixing ratio A1, A2, A3 and A4. Figure 2 (a) shows the signal curve of the pressure sensor after the detonation test of propane in pure air. It can be seen from the figure that when the ignition control system releases the ignition signal through the high-energy igniter, the mixed gas in the detonation tube is ignited, forming a detonation wave, and the flame surface moves downstream in the tube. The pressure in the tube is 0.35MPa, and the velocity v of the flame wave reaching the T4 sensor is 389.61m/s. The flame wave reaches the exit end face of the detonation tube wall after 1ms after ignition. At this time, a reflected shock wave is formed and propagates into the tube. When the reflected shock wave reaches T4 sensor, the peak pressure measured by T4 sensor is 0.74MPa, which is twice the pressure of the flame wave. Figure 2 (b), (c) and (d) all show the pressure signal curves in the detonation tube after propane combustion in oxygen-rich air. Under the condition of equal chemical equivalent ratio, the increase of oxygen content in the detonation tube greatly improves the detonation characteristics of propane. After the ignition head discharge, the pressure curves of T1, T2, T3 and T4 sensors rapidly rise to the maximum within 1-2μs after the arrival of the flame wave. The rise curves all have a very steep rise edge, which can be approximately seen as a straight line. By analyzing the difference of the time interval between the rise edge of different sensors, combined with the tube spacing between the sensor pairs, The propagation velocity v of the flame wave in the detonation tube can be obtained. In Figure 2 (b), (c) and (d), the propagation velocity v of the flame wave reaching the T4 sensor is 703.13m/s, 910.81m/s and 1800.85m/s successively. The pressure p in the pipe is 0.60 MPa, 0.82 MPa, 4.30 MPa, respectively. By comparing the above flame wave propagation velocity with the theoretical values of detonation performance parameters of the mixed gas in Table 1, it can be seen that the flame wave propagation velocity v and the pressure p in the tube of the three groups A1, A2 and A3 after the ignition of the mixed gas through the detonation tube are significantly different from the theoretical values, and neither of them reaches the detonation velocity of CJ. It can be concluded that the three groups of mixed gas only form a deflation flame surface in the detonation tube. There was no detonation wave; After the mixed gas of group A4 is ignited in the detonation tube, the flame wave velocity v and the pressure p in the tube reach the detonation velocity v_{CJ} of CJ, which can be regarded as forming a stable detonation state in the tube.

The ignition signal is input to the high-energy power supply through the ignition control system. It can be seen from the PCB sensor pressure signal curve that at 0.25s, the ignition head will release the electric spark to ignite the unignited gas in the detonation tube. The time for the flame wave to reach the sensor is calculated, combined with the position of each sensor in the distance from the ignition head to the detonation tube. The velocity v when the flame wave reaches the sensor position in the tube can be calculated, as shown in Figure 3. The curves A1, A2, A3 and A4 respectively correspond to the combustion velocity curves of mixed gas with different numbers in detonation tubes as a function of tube spacing. As can be seen from the figure, A1, A2 and A3 mixed gases do not reach the detonation velocity v_{CJ}-C_3H_8 after ignition, while A4 mixed gases reach the detonation velocity v_{CJ}-C_3H_8, indicating that a stable detonation state is formed. The flame propagates downstream along the detonation tube, and the velocity of flame wave gradually rises, reaching the maximum velocity of 436.89m/s, 720.00m/s, 1022.36m/s and 1851.85m/s respectively in the middle part of the detonation tube. Then the flame speed was reduced to 389.61m/s, 703.13m/s, 910.81m/s and 1800.85m/s. This is because the flame wave reaches over divergence after accelerating through the spoiler, at which time the flame propagation speed reaches the maximum, and then the flame velocity decreases to reach a stable flame propagation state.

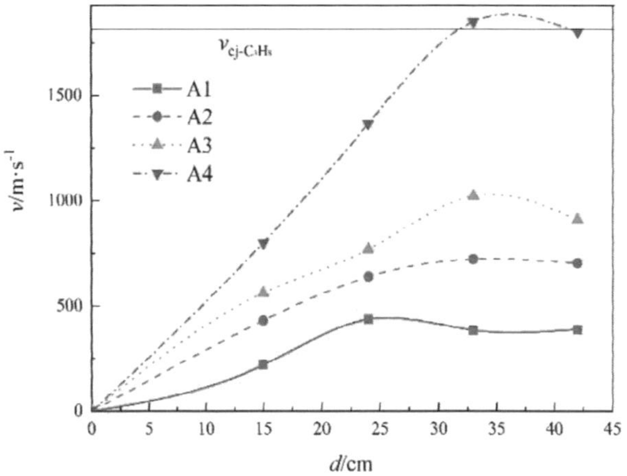

Figure 3. Flame propagation velocity variation curve with sensor position.

Figure 4 shows the flame propagation process of propane/air mixed gas in the detonation tube. The whole process is recorded by high-speed camera, in which the ratio of A1 mixed gas is propane/air equivalent ratio, and A2、A3 and A4 are propane/oxygen-rich air equivalent ratio. As can be seen from Figure 4 (a) (b)and (c), when propane is mixed with pure air or a small amount of oxygen-rich air for ignition, the ignition head of the detonation tube head releases low-energy electric spark to ignite the mixed gas at the front end of the intake section, generating flame wave and detonation pressure wave. The flame wave accelerates after passing through the turbulence plate and propagates to the rear section of the detonation tube, at which time detonation pressure wave appears blue. As the pressure wave propagates to the outlet of the detonation tube, the flame wave is still propagating in the middle of the tube. At this time, the pressure wave and the flame wave are decoupled, forming a yellow and blue reaction area. When the flame wave propagates to the tube port, the air moves from the tube port to the head of the detonation tube, and the flame gradually goes out. As can be seen from Figure 4 (d), when propane and a certain amount of oxygen-rich air are mixed and ignited, a bright flame wave is formed at the head. The flame wave quickly moves to the back end of the detonation tube after the acceleration action of the spoiler. The deuteration pressure wave and flame wave are decoupled at the back end of the spoiler, and the deuteration pressure wave quickly triggers the chemical reaction of unburned fuel in the back part of the tube, showing the reaction zone in blue. The flame wave follows the deflagration pressure wave to the rear section of the detonation tube. When the deflagration pressure wave reaches the end face of the detonation tube, a reflected shock wave is formed and propagates back to the tube, and coincides with the flame wave at a certain time, and the overlap section is clearly defined. When the reflected wave reaches the PCB dynamic pressure sensor, the pressure signal curve of the sensor rises abruptly, and the pressure size is twice that of the first flame wave. As the reflected shock wave propagates to the tube, the detonation tube presents a state of negative pressure, air enters the detonation tube, the flame in the tube gradually extinguishes, and the combustion pressure in the tube gradually decreases.

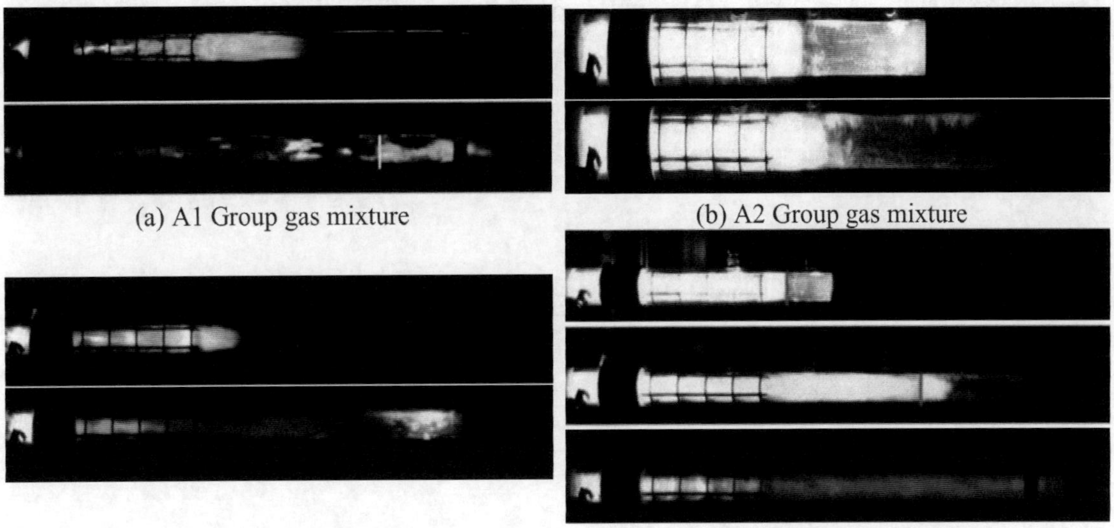

(a) A1 Group gas mixture (b) A2 Group gas mixture

(c) A3 Group gas mixture (d) A4 Group gas mixture

Figure 4. Propane/air mixed gas flame wave propagation process.

4. Conclusion

Based on the propane/air detonation characteristics, detonation tests of propane/air pulse detonation engine under different fuel oxygen content were carried out in this paper. In the experiment, detonation tests of propane/air/oxygen mixture gas under the appropriate chemical ratio were carried out to explore the effects of different oxygen-rich environment on detonation characteristics of propane/air pulse detonation engine. Analysis of experimental results in this chapter shows that:

(1) In the detonation experiment of propane/air/oxygen mixture with oxygen content of 20%, 25% and 30% in the fuel at the appropriate chemical ratio, the flame wave propagation velocity in the detonation tube is low, which fails to form a stable detonation wave, among which the fuel mixture with oxygen content of 20% has the worst detonation effect. The propagation velocity of the flame wave is 384.00 m/s.

(2) Under the appropriate propane/air chemical ratio, the propane mixture requires a higher ignition temperature in the detonation tube, while the spark plug energy is low, and the propane/air mixture gas ignition is difficult to achieve stable combustion.

(3) The pressure signal of the PCB dynamic pressure sensor measured during the detonation experiment of the propane/air/oxygen mixture with oxygen content of 35% in the fuel at the exact chemical ratio has a steep rising edge. The detonation velocity and pressure obtained through the pressure signal analysis are both greater than the theoretical CJ detonation velocity, and the detonation state is successfully reached.

References

[1] Movileanu C , Mitu M and Giurcan V 2020 *J. Fuel*, **274** 117836.

[2] Tobing R L and Ridwan R 2011 *J. Jaids Journal of Acquired Immune Deficiency Syndromes*, **46(46)** 248-52.

[3] Takeshima N , Ozawa K and Tsuboi N 2020 *J. Shock Waves*, **30(1)** 1-16.

[4] Frolov S M, Smetanyuk V A and Shamshin I O 2020 *J. Applied Thermal Engineering*, **183** 1-116.

[5] Ting J , Bussing T and Hinkey J 2013 *Conf. Joint Propulsion Conference & Exhibit*, p 3154.

[6] Li C M , Chung K M and Lai W H 2006 *Conf. AIAA Aerospace Sciences Meeting & Exhibit*. p 554.

[7] Yu H and Chen Z 2015 *J. Combustion and Flame*, 2015, **162(11)** 4102-4111.

[8] Lu F K , Meyers J M and Wilson D R 2003 *J. AIAA Paper*, p 171.

Body effect on inflow distortion in civil engine intake under crosswind condition

Yan Li*, Yingchun Chen, Pan Cheng and Dongyun Zhang

Shanghai Aircraft Design and Research Institute, Shanghai, China

* Corresponding author's email: liyan61329@126.com

Abstract. The effect of body on the inlet aerodynamic performance of distortion under crosswind condition for a high pass ratio turbofan engine was investigated. The work is carried out by using computational fluid dynamics on civil nacelle and body. The investigation suggests that: due to the shelter of the body, the inlet distortion on the windward side and leeward side is very different under crosswind conditions. The nacelle on the windward side mainly takes in air from the side while the engine on the leeward side takes in air from above and below. The existence of body makes it easier to produce fuselage vortex and tail vortex near the nacelle. The body has an adverse impact on the intake of the engine on the leeward side and improves the requirement of the crosswind resistance of the upper section of the nacelle.

1. Introduction

The primary role of the air intake is to provide the fan with adequate flow uniformity over a wide range of operating conditions. Modern civil turbofan engines operate at high bypass ratios and have larger diameters, thus requiring shorter lengths to compensate for increased intake drag. Shorter air intakes reduce the ability to distribute evenly, which results in easier separation of airflow over the cabin compared to conventional air intakes.

In order to meet the needs of the integrated design and analysis of the airframe and the engine, CFD and wind tunnel tests were carried out in the process of civil aircraft. Recently, CFD tools have also been applied to simulate inlet vortices with relative success. Trapp et al.[1] performed RANS simulations to evaluate vortex behavior around the nacelle under crosswind conditions, and validated the numerical results with measured data and analytical solutions. Cao et al. [2] conducted a numerical study on the interaction between airflow separation on the air intake and the fan at high incidence. Zantopp[3] conducted an experimental study to study the formation of ground vortex from air inlets under crosswind conditions. Kuen[4] studied the effect of inlet disturbances caused by crosswinds on the operation of fan blades. Verm[5] studied the effect of total pressure distortion on single-stage compressors and low bypass ratio fans. Mauro[6] investigated the conditions under which distortion occurs in the intake air of a civil aircraft engine, investigating two sources of distortion: ground vortex intake and high angle of incidence flight conditions. Menezes[7] analyzed the root cause of the EMB-145 engine core speed being higher in the right engine than the left engine through CFD analysis and aircraft testing, considering the crosswind effect. The effect of bodywork on intake tract distortion under crosswind conditions has been rarely studied before.

This paper reports the results of a numerical study of fuselage and nacelle configurations with radial total pressure deformation. This paper is organized as follows. The numerical model and benchmark test cases are first introduced. Afterwards, the numerical method used here is validated by comparison with known test data. Finally, the effect of the fuselage on the intake distortion of the engine nacelle under crosswind is investigated.

2. Physical model

A typical civil aircraft engine power nacelle with a large bypass ratio is selected as the research object to study the influence of the airframe on the crosswind intake distortion of the civil aircraft intake. The power nacelle model is shown in Figure 1. The nacelle includes an air inlet, an air intake cone, a fan cover, an external outlet and an inner outlet. Considering the effect of ground effect, the ratio of intake port height H to nacelle diameter D is H/D=0.93.

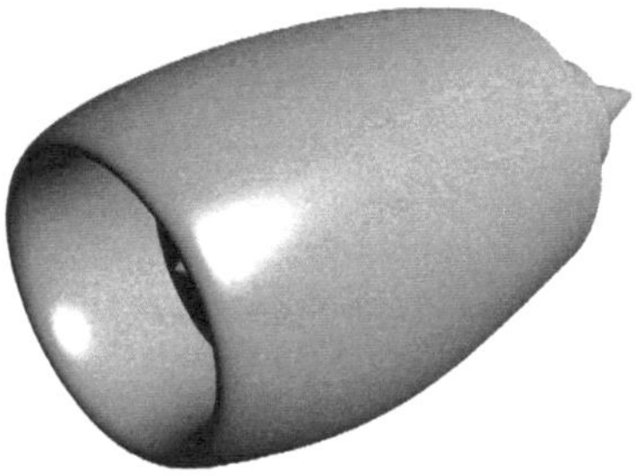

Figure 1. Schematic diagram of the nacelle.

3. Calculation method and calculation conditions

The ANSYS ICEM software is used to generate a structured computational grid for the full 3D flow field of the fuselage and nacelle. Fig. 2 shows the schematic diagram of the computational grid for the fuselage and nacelle surfaces. In order to ensure the accuracy of the calculation results and capture the flow process of jet and mixing well, the distance from the far-field boundary of the computational domain to the nacelle is set to 250 times the equivalent diameter of the nacelle. In order to accurately simulate the complex flow in the boundary layer, the grid near the wall of the nacelle is refined. The grid height of the first layer is about 0.01 mm, and the near wall is about $y+ \approx 1$. In addition, the meshes of the inlet, lip, inner and outer culverts, and the areas near the ground where the flow field parameters change violently are properly refined. The total number of three-dimensional computing grids in the entire flow field is about 38 million.

The software used for numerical simulation is ANSYS CFX, and CFX uses lattice-based finite volume method to discretize the Reynolds-averaged Navier-Stokes equations. The computational domain to be solved is first divided into different types of mesh elements, and then control volumes are built around each mesh node. The flux of the control volume is calculated by integrating points on the face between the control volumes. The time discretization adopts the fully implicit time advancing format, and the spatial discretization adopts the high-order upwind format. The Reynolds stress in the momentum equation is handled by the two-equation turbulence model SST combined with the automatic wall function. The model uses the k-ω model in the near-wall region and the k-ε model in the far-wall region. In the near-wall region, it not only has the characteristics of high calculation accuracy of the k-ω model, but also overcomes the k-ω model in the far-wall region. The model is sensitive to incoming flow ω.

In the flow field calculation, the residuals of the whole flow field and the inlet and outlet flow are monitored, and the convergence is based on the fact that each residual index decreases by 4 orders of magnitude or the residuals do not decrease after continuing iterative iterations and the inlet and outlet flow remains stable.

The boundary conditions are set as follows: the far-field inlet boundary condition adopts the velocity inlet condition; the far-field outlet boundary condition adopts the pressure boundary condition; the calculation parameters are given the parameters such as free flow static temperature, static pressure, and flow velocity. Crosswind conditions are imposed in the calculation, and the direction of the crosswind is from right to left (driver's perspective, the right side is the windward side). The inlet boundary conditions of the intake duct adopt the pressure boundary conditions, which are continuously adjusted with the iterative process to match the flow rate of the fan surface; the boundary conditions of the outer duct inlet adopts the total temperature and total pressure boundary conditions; the boundary conditions of the inner duct inlet adopt the total temperature and total pressure boundary conditions; the nacelle object surface boundary condition applies the solid-wall no-slip adiabatic condition. The ground boundary condition is set to the no-slip wall boundary condition.

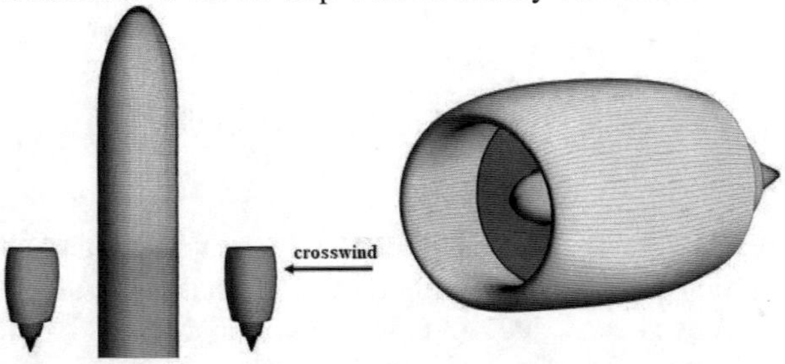

Figure 2. Schematic diagram of surface meshes for fuselage and nacelle calculation.

4. Calculation results and discussion

After the text edit has been completed, the paper is ready for the template. Duplicate the template file by using the Save As command, and use the naming convention prescribed by your conference for the name of your paper. In this newly created file, highlight all of the contents and import your prepared text file. You are now ready to style your paper; use the scroll down window on the left of the MS Word Formatting toolbar.

4.1. Verification

In order to verify the ability of CFD software to solve the flow problems related to the power nacelle, the experimental data of the axisymmetric turbofan power simulator (TPS, turbine powered simulator) wind tunnel of Japan Institute of Aeronautics and Astronautics (JAXA) was selected to verify the numerical example. . The calculation adopts the working condition of calculation number 13 (corresponding to the experimental number 516) in the literature [8], the calculation Mach number is 0.801, the Reynolds number is 2 million, the temperature is 315.74K, the attack angle is 0 ° , the sideslip angle is 0 ° , and the mass flow coefficient is 0.31678. The ratio of the total pressure of the external nozzle to the static pressure of the incoming flow is 0.92978, the ratio of the total temperature to the static temperature of the incoming flow is 1.01198, the ratio of the total pressure of the internal nozzle to the static pressure of the incoming flow is 0.68668, and the ratio of the total temperature to the static temperature of the incoming flow is 0.99310 . Fig. 4 shows the distribution curves of the calculated and experimental results of the pressure coefficients on the nacelle fan cover and the inner fairing. It can be seen from the figure that the calculated pressure coefficients in the inner and outer flow areas of the nacelle are basically consistent with the experimental values. It can be considered that the numerical method, grid model and the treatment measures of the inlet and exhaust boundary conditions used in this paper can be more accurate. Simulate the flow field characteristics of the nacelle intake, shroud and exhaust system.

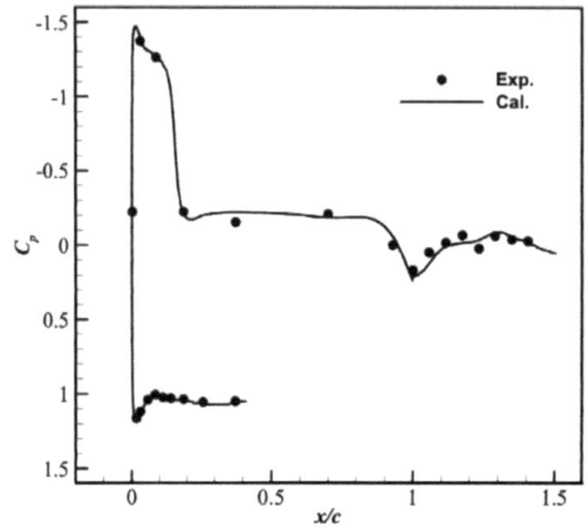

Figure 3. Comparison of nacelle pressure coefficient and test value.

4.2. Streamlines

In this paper, the typical ground take-off state is used to analyze the air intake distortion characteristics of the nacelle. The calculation parameters are: height 0 m, static pressure 101325 Pa, and static temperature 288.15K. The reduced flow rate of engine power is about 533 kg/s, the angle of attack is 0 °, and the side slip angle is 90 °.

The flow diagram of the ground crosswind condition when the crosswind speed is 30 knots is shown in Figure 4. The part of the side flow near the engine lip first enters the windward side engine, and the part farther from the engine lip bypasses the fuselage , into the leeward side engine. Due to the shielding of the fuselage, the air intake of the nacelles on the windward side and the leeward side under crosswind conditions is very different. A nacelle on the leeward side has a very different intake in crosswind conditions than a nacelle alone.

Figure 5 shows the intake range of the nacelles on the windward side and the leeward side. The leeward side nacelle absorbs the airflow from the front and farther, and the intake range is wider and is more affected by the uneven airflow around it. And its suction is affected by both the ground and the fuselage. It is easier to generate fuselage vortices.

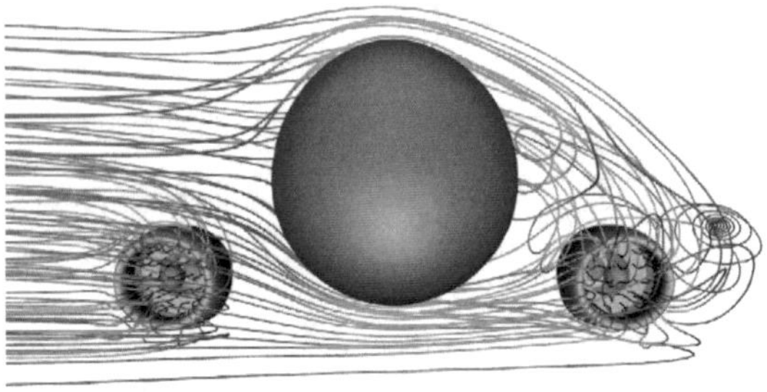

Figure 4. Flow diagram of ground crosswind condition.

(a) Nacelle on the windward side

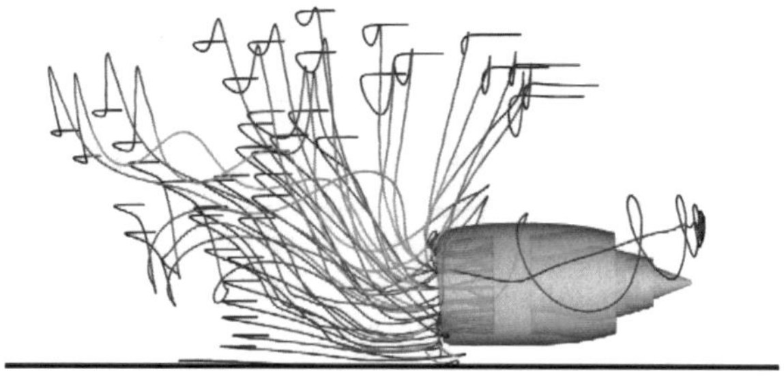

(b) Nacelle on the leeward side

Figure 5. Schematic diagram of nacelles air intake.

Generally, in order to reduce drag, the top section of the nacelle is thinner, and the crosswind section is the main windward side of the crosswind, and the design is thicker, but for the leeward side engine, the top section is the main windward side. The thin profile of the top is not good for the air intake distortion under crosswind conditions, and the experiment and CFD simulation of the single nacelle cannot reflect this effect. The influence of the fuselage on the intake of the leeward side engine cannot be ignored.

4.3. Separation

According to the overall design and safe operation requirements, the air intake should meet the engine air intake quality requirements at least at 20 knots of crosswind under ground conditions. In this paper, the influence of crosswind speed on intake distortion is firstly studied under the certified engine operating conditions, and the range of the analyzed crosswind speed is 10-30 knots. For the convenience of explanation, the following describes the size of the crosswind speed in conjunction with the imperial unit (knot) commonly used in the evaluation of the crosswind speed of civil aircraft.
Figure 6 shows the distribution of total pressure recovery at the outlet of the left and right air ducts, that is, the fan section when the crosswind speeds are 15 knots, 20 knots, 25 knots and 30 knots, respectively. The red area in the figure is a high value and the blue The color area is the low value, the left side is the leeward side, and the right side is the windward side. It can be seen from Figure 6 that the left and right engine nacelles have low total pressure areas near the wall, which is due to the decrease in total pressure caused by the loss of viscosity due to the airflow. The distribution of total pressure at the outlet of the right engine (windward side engine) is basically similar, and the low total pressure area causing the increase of the distortion index is mainly distributed near the wall and the ground vortex area. The size and intensity of the ground vortex area in the low pressure area near the wall increase with the increase of the crosswind speed.
The left engine is used as the wind-driven side engine. When the crosswind speed is less than 25 knots, the ground vortex phenomenon is not as obvious as that of the right engine. At the crosswind speed of 25 knots, an obvious low total pressure area appears on the top of the nacelle. At the crosswind speed of 30 knots, the The low total pressure area becomes larger. The distortion index will increase significantly at this time. This is due to the thin profile at the top of the nacelle, where the airflow separates as it accelerates.

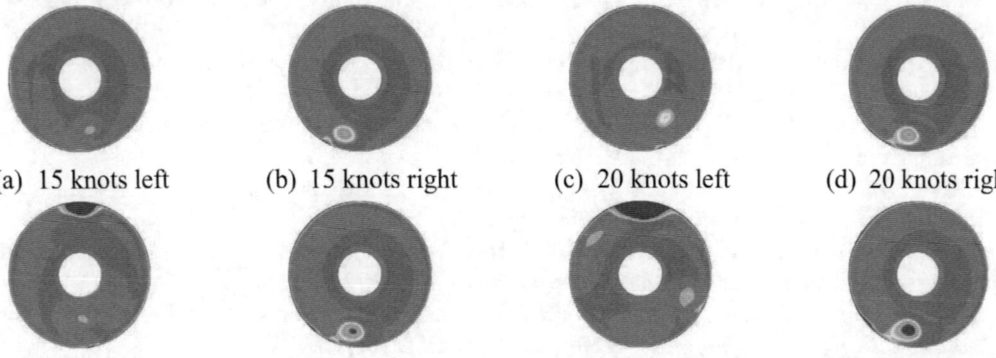

(a) 15 knots left (b) 15 knots right (c) 20 knots left (d) 20 knots right
(e) 25knots left (f) 25 knots right (g) 30 knots left (h) 30 knots right

Figure 6. Distribution of total pressure recovery on exit of inlet.

4.4. Vortex

Figure 7 shows the recovery distribution of the total pressure at the intake port and the vorticity isosurface when the crosswind speed is 30 knots, and the isosurface shows the area where the force has vortices. It can be seen from the figure that the ground vortex suction phenomenon is obvious, resulting in an increase in the stagnant pressure loss on the fan surface. The fuselage vortex shedding after flowing through the nose may be sucked into the left engine (by the wind side engine), causing separation and aggravating total pressure distortion. The existence of the fuselage makes the flow field environment where the engine on the wind-driven side is located more complicated, which adversely affects the air intake quality of the wind-driven side engine.

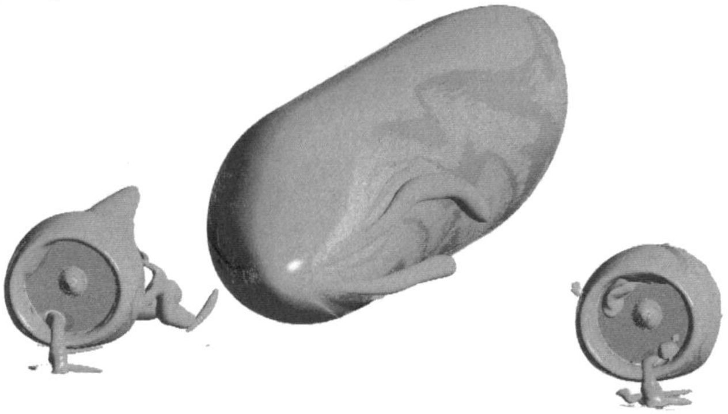

Figure 7. Space vorticity isosurface.

5. Conclusion

The influence of the fuselage on the air intake is reflected in the change of the main air intake channel (air intake direction) of the engine on the wind side under the crosswind condition, the expansion of the incoming air flow range, the generation of fuselage vortices, and the increase in the incoming flow. The non-uniformity of the air intake is ultimately reflected in the increase of the intake distortion index, which has an adverse effect on the intake air quality. Therefore, the influence of the fuselage must be considered when designing the nacelle air intake.

Acknowledgments

The authors gratefully acknowledge National Key R&D Program of China (Project No. 2020YFA0712000) for funding this work.

References

[1] Trapp L and Motta R 2010 Crosswind Effects on Engine Inlets: The Inlet Vortex *Aircraft* 47(2), pp 577-590
[2] Cao T, Vadlamani N, Tucker P, Smith A, Slaby M and Sheaf C 2016 Fan-Intake Interaction under High Incidence *ASME* paper No GT2016-56561
[3] Zantopp S, MacManus D and Murphy J 2010 Computational and Experimental Study of Intake Ground Vortices *Aeronaut* 114(1162) pp 769-784
[4] Kuen B, Mark W and Mehdi Vahdati 2019 Effects of inlet disturbances on fan stability *Journal of Engineering for Gas Turbines and Power*, 141(5)
[5] Verma V, Singh G and Pradeep A 2020 The effect of inlet distortion on low bypass ratio turbofan engines *Proceedings of the Institution of Mechanical Engineers* 234(8)
[6] Mauro C, Feng W and Luca M 2017 Low Frequency Distortion in Civil Aero-engine Intake, *Journal of Engineering for Gas Turbines and Power* 139(4)
[7] Menezes S, Frederick B and Robert P 2006 Crosswind Effects in Rear-Mounted Aircraft Engine Installation *SAE TECHNICAL PAPER SERIES* 2006-01-2882
[8] Naoki H, Keisuke A and Katsuya I 1989 3D-Euler flow analysis of fanjet engine and turbine powered simulator with experimental comparison in transonic speed *AlAA 20th Fluid Dynamics*

Numerical analysis of flow and heat transfer characteristics of supercritical endothermic hydrocarbon fuel in taper tube

Jiajun Zhao[1, 2, 3], Hui Gao[1, 2, 3,*], Chen Zhang[1, 2, 3] and Dongsheng Wen[1, 2, 3, 4]

[1] School of Aeronautic Science and Engineering, Beihang University, Beijing 100191, PR China

[2] Zhongfa Aviation Institute, Beihang University, Hangzhou, Zhejiang 311115, PR China

[3] Ningbo Institute of technology, Beihang University, Ningbo 315100, PR China

[4] Institute of Thermodynamics, Technical University of Munich, Munich, Germany

* Corresponding author's email: h.gao@buaa.edu.cn

Abstract. To enhance the cooling performance of the scramjet regenerative cooling system, a novel taper tube is proposed. Under supercritical pressure, the flow characteristics and heat transfer characteristics of endothermic hydrocarbon fuel (EHF) in taper tube, narrow cylindrical tube, and wide cylindrical tube are numerical investigated and compared. The results of the investigation of flow and heat transfer processes in the above three tubes indicate that the taper tube effectively facilitates the pyrolysis of *n*-decane and the *n*-decane conversion is increased by 3.94%, compared with narrow cylindrical tube. Compared with wide cylindrical tube, the taper tube enhances cooling effect and the wall temperature near the outlet is reduced by 50.16 K. The taper tube contributes to the cooling structure optimization and promotion of regenerative cooling system performance.

1. Introduction

Scramjet is a promising propulsion device for hypersonic vehicle, which becomes a research focus in the propulsion field. However, the combustion chamber of scramjet faces severe thermal environment, which brings great challenges to the existing thermal protection technology and materials. The regenerative cooling method, which uses endothermic hydrocarbon fuel as the cooling system coolant, an effective and realizable active thermal protection technology, is one of the most promising technologies [1, 2].

However, the regenerative cooling system application is restricted by a shortage of cooling capabilities. Hence, many optimized structures of cooling channel have been designed to enhance regenerative cooling performance, and the detailed endothermic hydrocarbon fuel flow characteristics and heat transfer characteristics in the optimized cooling tubes have been investigated via computational fluid dynamic (CFD) simulation [3, 4].

The cross-section optimization of regenerative cooling tubes, which significantly affects the cooling performance of regenerative cooling systems, has received extensive attention [5]. Li *et al.* [6] researched the thermal cracking and carbon deposition phenomenon of EHF in the regenerative cooling tubes with different cross-section but same cross-sectional area, including rectangular, square, and circular cross-sections, based on electric heating experiment and CFD simulation. However, the optimization methods to improve pyrolysis and carbon deposition phenomenon were not proposed.

Jiang *et al.* [7] proposed a triangular shape cross section channel, which improves heat transfer performance and flow distribution in parallel-channel-configuration. But the pyrolysis of *n*-decane in single triangular shape channel should be further discussed. Li *et al.* [8] designed a spiral rectangular channel to enhance turbulence intensity and induce spiral flow, which enhanced heat transfer effects. However, in their work the pyrolysis of *n*-decane, an important factor of regenerative cooling system, was neglected.

The optimization of regenerative cooling tube cross-section is an effective model to improve the cooling performance and heat transfer property and the EHF chemical heat sink utilization. To fully utilize EHF chemical heat sink and reduce temperature of the wall near the tube outlet, a taper tube with a cross section that shrinks along the flow direction is proposed. The cooling performance of taper tube, narrow cylindrical tube, and wide cylindrical tube, including pyrolysis of EHF, flow characteristics and heat transfer characteristics in the tubes, is comparative investigated via CFD simulation. This study contributes to the structural optimization and the performance promotion for the regenerative cooling tubes.

2. Numerical methods

2.1. Geometry of cooling tubes and boundary conditions of model

In this study, three 2D numerical models which is shown in figure 1, are established to simulate the flow characteristic and heat transfer characteristic of EHF. According to the actual structure of the scramjet, the heating section length is set to be 1000 mm [9]. Two adiabatic sections with the length of 100 mm are established to ensure the flow of fluid in the cooling tube model fully developed and strengthen stability of simulation, respectively. Each of tube is designed with a total radius of 1.5 mm, comprising the fluid model zone and the solid model zone. The inlet radius and outlet radius of taper tube (Tube-1) are 1 mm and 0.5 mm, respectively. For the adiabatic section, the fluid domain radius is constant. The fluid domain of taper tube shrinks smoothly along the direction of fluid flow in the heating section. The fluid domain radii of Tube-2 and Tube-3 are respectively 0.5 mm and 1 mm. The constant heat flux is defined on the heated surface. The mass flow tube inlet and pressure tube outlet are defined, and the tube inlet flow rate of mass in the three tubes is 3.0 g s^{-1}. The detailed values of boundary conditions are shown in table 1.

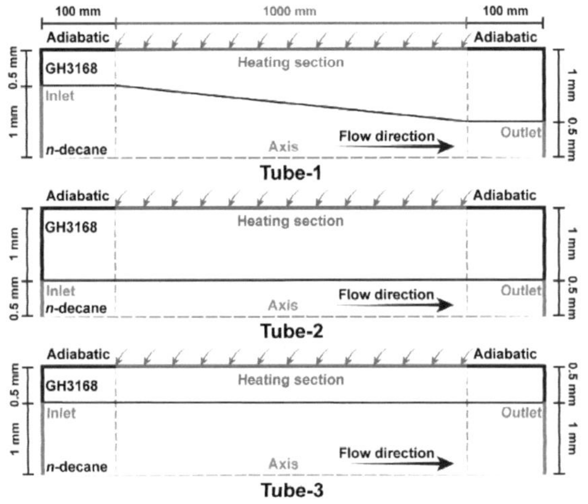

Figure 1. Three 2D axisymmetric numerical models' schematic diagram.

Table 1. Detailed values of boundary conditions

Inlet flow rate of mass, g s^{-1}	Temperature of inlet, K	Operating pressure, MPa	Heat flux of heating section, W m^{-2}
3.0	373.15	3.45	800000

2.2. Governing equations and solution methods

The steady state flow and heat transfer characteristics of EHF are investigated with the commercial software ANSYS Fluent. The SIMPLEC algorithm is applied to resolve the coupling of pressure and velocity. The conservation equation of mass, conservation equation of momentum, conservation equation of energy, and conservation equation of species transport are listed as following:

Conservation equation of mass:

$$\nabla \cdot (\rho \vec{u}) = 0 \tag{1}$$

Conservation equation of momentum:

$$\nabla \cdot (\rho \vec{u}\vec{u}) = -\nabla p + \nabla \vec{\tau}_{eff} + \rho \vec{g} \tag{2}$$

Conservation equation of energy in fluid zone:

$$\nabla \cdot (\rho \vec{u} e_t + p\vec{u}) = \nabla \cdot (\lambda_{eff} \nabla T) \tag{3}$$

The conservation equation of species transport, which is used to describe the thermal cracking of EHF:

$$\nabla \cdot (\rho Y_i \vec{u}) = -\nabla \cdot \left(\rho Y_i \vec{u}_{d,i}\right) + S_i \tag{4}$$

where the mass source term S_i in the conservation equation of species transport is expressed as:

$$S_i = \omega_i M_{wi} \tag{5}$$

The nickel-base superalloy (GH3128) is selected as wall metal material. In the solid domain, the conservation equation of energy is expressed as:

$$\nabla \cdot (\lambda \nabla T) = 0 \tag{6}$$

The SST k-ω turbulence model is used to describe the fluid turbulent flow in three tubes, which is a low speed and incompressible flow constrained by walls.

$$\nabla \cdot (\rho \vec{u} k) = \nabla \cdot [(\mu + \mu_t/\sigma_k)\nabla k] + G_k - Y_k \tag{7}$$

$$\nabla \cdot (\rho \vec{u} \omega) = \nabla \cdot [(\mu + \mu_t/\sigma_\omega)\nabla \omega] + G_\omega - Y_\omega + D_\omega \tag{8}$$

2.3. Pyrolysis mechanism

The model fuel, *n*-decane, is chosen in this work, which has similar thermo-physical properties to aviation kerosene. The PPD model (proportional product distribution model of *n*-decane), which is a one-step surrogate thermal cracking reaction model, is selected to describe the pyrolysis in the regenerative cooling tube [10].

The thermal cracking reaction rate of *n*-decane is calculated by:

$$-\frac{d[C_{10}H_{22}]}{dt} = k_{py}[C_{10}H_{22}] \tag{9}$$

$$k_{py} = A_{py} e^{-E_a/(RT)} \tag{10}$$

In the simulation model of regenerative tubes, the pre-exponential factor (A_{py}) detail value is set to 1.6×1015 s^{-1} and the activation energy (E_a) detail value is set to 2.63655×10^8 J kg^{-1} mol^{-1}.

Due to the transcritical process and pyrolysis, the EHF thermos-physical properties in the regenerative cooling tubes change drastically. To descript the thermos-physical properties accurately, the SUPERTRAPP Software from NIST, which is widely applied in the former research, is used in this work. And both the *n*-decane thermo-physical properties and the thermo-physical properties of mixture of its pyrolytic products are calculated by SUPERTRAPP.

3. Results and discussion

3.1. Validation of regenerative cooling tube model

To validate the reaction model, a 2D axisymmetric numerical model is established. The geometry of regenerative cooling tubes and boundary conditions of regenerative cooling tube model are consistent with those in the experiment research. A 2D structural mesh, generated by ANSYS ICEM, is used for the validation after the mesh independence test being made [26]. The axial velocity of fluid in the tube and *n*-decane mass fraction are chosen for the comparison between numerical simulation and experiment. The maximum error is less than 10%.

Three meshes of taper tube are established for mesh independence verification. The detailed information of three meshes is shown in table 2. The boundary conditions of regenerative cooling tube in this test are consistent with those shown in table 1. The axial heating wall temperature distributions of three meshes are compared. From figure 2, the maximum error of mesh independence verification is less than 10%. Hence, mesh-2 is chosen in this research. Moreover, the value of *y*+ is less than 1.0, which ensure the first layer of the mesh falls into the viscus sublayer.

Table 2. Detailed mesh information of three taper tube meshes.

Mesh number	Mesh information					
	Sectional division in fluid zone	Mesh initial Height	Mesh growth Rate	Sectional division in solid zone	Longitude division number	Total number
Mesh-1	41	1 μm	1.05	11	2400	120000
Mesh-2	51	1 μm	1.05	11	2400	144000
Mesh-3	61	1 μm	1.05	11	2400	168000

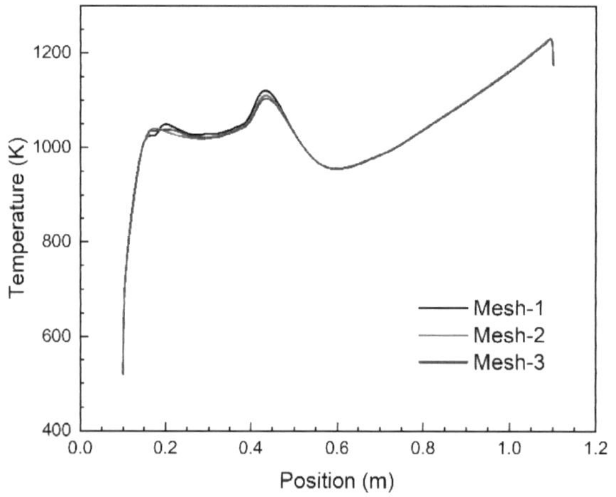

Figure 2. Heating wall temperature profile of three meshes.

3.2. Effect of taper tube on velocity characteristics

When the inlet flow rate of mass is constant, the fluid velocity is significantly influenced by tube cross-sectional area. The fluid velocity along the centerline of three tubes is illustrated in figure 3. Due to the smaller cross-sectional area, in main segment, fluid flows more quickly in tube-2 compared to that in tube-1, which is beneficial to heat fluid and enhance *n*-decane cracking reaction. In the initial segment, fluid velocity of tube-1 and tube-3 are nearly equivalent. As the cross section of tube-1 shrinks, fluid flows more rapidly than that in tube-3, which contribute to cooling the metal wall. In comparison to tube-3, tube-1's outlet velocity rises by 155.13%. In addition, fluid is significantly accelerated along flow direction, which is thermal acceleration phenomenon caused by heating and pyrolysis. And in the end segment, fluid flows more quickly in tube-1 than tube-2, due to more significant acceleration phenomenon carried by more dramatic pyrolysis.

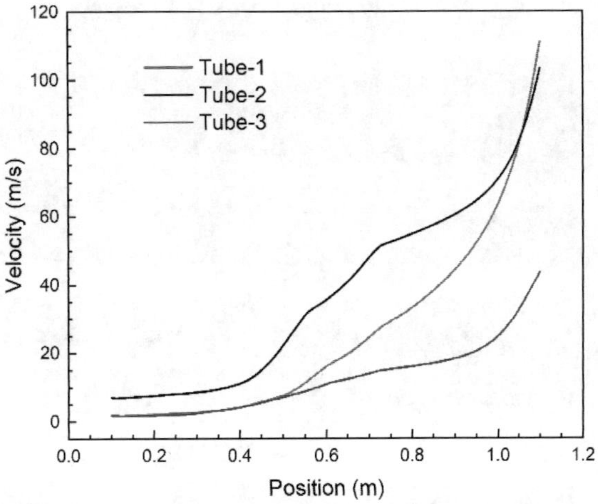

Figure 3. Fluid velocity profile of centerline in three tubes.

3.3. Effect of taper tube on temperature characteristics

Figure 4 shows that, the wall temperature of tube-2, benefited by faster coolant flow rate, is lower than that of other tubes, especially in initial segment. Accompanied by the shrinkage of cross section, the wall temperature of tube-1 closes to that of tube-2 and it's a decrease of 50.16 K compared to that of tube-3. There is a soaring temperature in the wall of tube-3, and this phenomenon is weakened in tube-1.

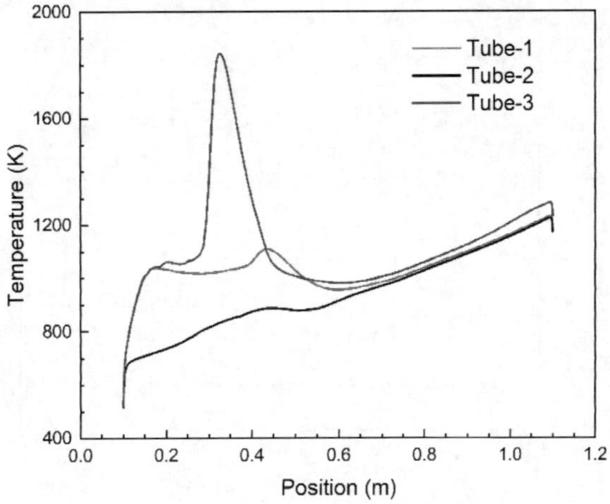

Figure 4. Wall temperature profile of three tubes.

Figure 5 shows that, the EHF close to the solid wall gets heated quickly in the tube-1 and tube-3, which contributes to the *n*-decane thermal cracking reaction. However, there is a concentration of high temperature fluid in tube-3, which may cause thermal damage to the wall metal material. In tube-1, this phenomenon is so weak that it can be ignored. There is significant thermal stratification in tube initial segment. The main fluid is not fully heated, which degrades cooling performance and weakens pyrolysis of *n*-decane. At the rear, the fluid is sufficiently heated, and this phenomenon is diminished.

Journal of Physics: Conference Series **2633** (2023) 012011 doi:10.1088/1742-6596/2633/1/012011

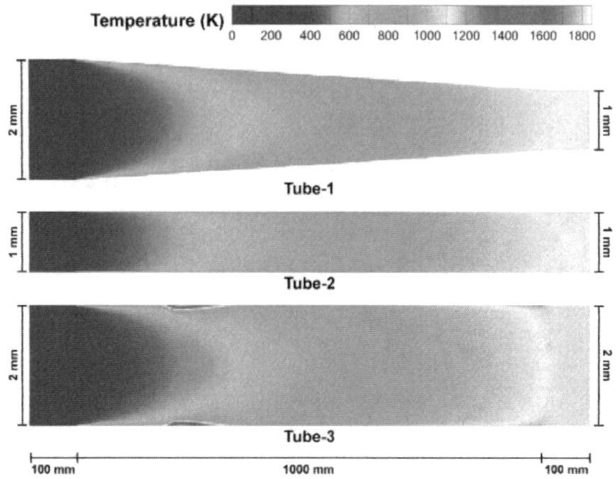

Figure 5. Temperature contours of fluid in three tubes.

3.4. Effect of taper tube on temperature characteristics

For the regenerative cooling system, the EHF should be as thoroughly cracked as possible to fully use the EHF chemical heat sink and enhance its combustion performance. From figure 6, the pyrolysis of *n*-decane is greatly enhanced by fairly high fluid temperature. Comparing between tube-1 and tube-2, the thermal cracking reaction conversion of *n*-decane inside tube-1 is increased, due to extended mainstream flow resident time accompanied by elevated fluid temperature. The total conversions in three tubes are shown in table 3.

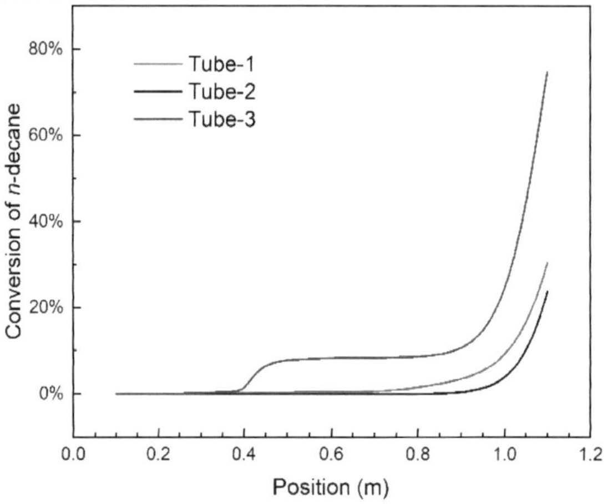

Figure 6. Conversion axial profile of *n*-decane in three tubes.

Table 3. Total conversion of *n*-decane in three tubes.

Tube number	Total conversion of *n*-decane
Tube-1	60.47%
Tube-2	56.53%
Tube-3	98.88%

Due to the promotion of pyrolysis reaction by higher temperature and longer residence time, as exhibited in figure 7, the *n*-decane in tube-1 begins to crack earlier than in tube-2. In addition, there is a concentration of reaction product near the wall, which hinders heat transfer between main fluid and metal wall carried by low thermal conductivity of reaction product.

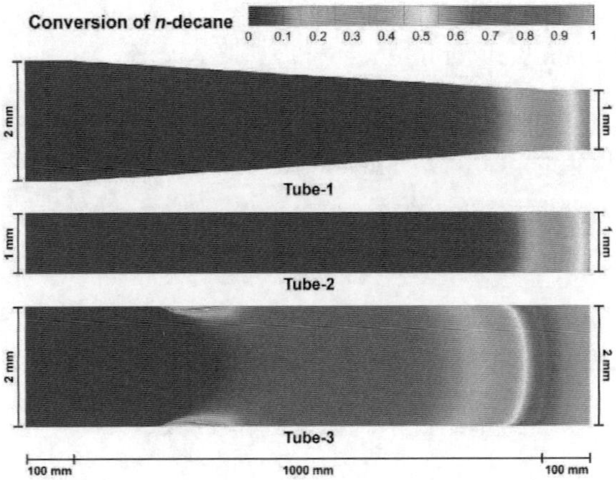

Figure 7. Conversion contours of *n*-decane in three tubes.

4. Results and discussion

To improve the cooling capacity and heat transfer performance of active regenerative cooling system, a novel taper tube is proposed. The flow characteristic and heat transfer characteristic of *n*-decane in the taper tube are numerically researched and compared with two traditional straight cylindrical tubes. The results indicate that the taper tube effectively slows down the flow and raises the temperature of the fluid in initial segment, which facilitates the thermal cracking reaction of *n*-decane and contributes to the earlier onset of the cracking reaction. Compared with narrow cylindrical tube (Tube-2), the total conversion in taper tube is increased by 3.94%. As the cross section shrinks, the fluid flow is accelerated, resulting in lower wall temperature. Compared with wide cylindrical tube (Tube-3), the wall temperature of taper tube near the outlet is reduced by 50.16 K. This work may provide reference for structural design of regenerative cooling tube.

Acknowledgments

The authors acknowledge financial support from Natural Science Foundation of Zhejiang Province (NO. KYZ1121004).

References

[1] E. T. Curran, Scramjet Engines: The First Forty Years, Journal of Propulsion and Power 17 (2001) 1138-1148.

[2] S. Zhang, X. Li, J. Zuo, J. Qin, K. Cheng, Y. Feng and W. Bao, Research progress on active thermal protection for hypersonic vehicles, Progress in Aerospace Sciences 119 (2020).

[3] S. Luo, D. Xu, J. Song and J. Liu, A review of regenerative cooling technologies for scramjets, Applied Thermal Engineering 190 (2021).

[4] C. Zhang, H. Gao, J. Zhao, J. Zhou, X. Ma and D. Wen, Dynamic coking simulation of supercritical *n*-decane in circular tubes, Fuel 331 (2023) 125859.

[5] S. Zhang, J. Qin, K. Xie, Y. Feng and W. Bao, Thermal behavior inside scramjet cooling channels at different channel aspect ratios, Journal of Propulsion and Power 32 (2016) 57-70.

[6] Fuqiang, Zaizheng, Jing, Kai, Wang, Li, Zhang, Xiangwen, Liu and Guozhu, Thermal Cracking of Endothermic Hydrocarbon Fuel in Regenerative Cooling Channels with Different Geometric Structures, Energy & Fuels (2018).

[7] Y. Jiang, Y. Xu, J. Qin, S. Zhang, K. Chetehouna, N. Gascoin and W. Bao, The flow rate distribution of hydrocarbon fuel in parallel channels with different cross section shapes, Applied Thermal Engineering 137 (2018) 173-183.

[8] N. Li, Z. Huang, Y. Bu and Y. Guan, Effects of channel geometries and acceleration on heat transfer of hydrocarbon fuel, Journal of Thermophysics and Heat Transfer 34 (2020) 570-578.

[9] C. Zhang, H. Gao, J. Zhao, J. Zhou and D. Wen, Cooling performance optimization of supercritical hydrocarbon fuel via bi-directional flow pattern, Applied Thermal Engineering 212 (2022) 118563.

[10] T. A. Ward, J. S. Ervin, S. Zabarnick and L. Shafer, Pressure Effects on Flowing Mildly-Cracked *n*-decane, Journal of Propulsion and Power 21 (2005) 344-355.

Numerical simulation of ply angle deviation and thermal deformation behavior of CFRP reflector

Wenjian Jiang*, Hengkun Jiang, Tonglong Huo and Yun Liang

Institute of Space Antenna Technology, Academy of Space Information Systems, Xi'an, 710100, People's Republic of China

* Corresponding author's email: hellojian88@126.com

Abstract. As we know, the ply angle deviation of Carbon Fiber Reinforced Polymer(CFRP) reflector is difficult to obtain accurately in engineering applications, it will make thermal deformation of reflector unpredictable due to the discreteness of ply angle. In this paper, based on constitutive relationships between generalized internal force and strain of laminated composite, a novel ply angle deviation model based on normal distribution is proposed. Furthermore, an orthogonal test was designed and significance analysis were carried out by means of a 2.6m aperture reflector with polygonal back structure. Assumed that the ply angle deviation obeys normal distribution, a number of samples were further randomly generated, and the sample mean and standard deviation of the thermal deformation of reflector were calculated. According to $\chi 2$-test of goodness of fit, the thermal deformation of reflector still follows normal distribution. Finally, an example was employed to verify the validity and effectiveness of the proposed method. The results reach a good agreement with the experimental data, thus it is capable of providing a promising prospect for engineering practice.

1. Introduction

With the rapid development of high-throughput communication satellites (HTS), extremely high thermal stability requirements for antennas in space environment will be needed to achieve a high gain, low sidelobe and high C/I(Carrier to Interference ratio) goal[1-2]. Because of simple structure, excellent weight, low cost and good structural adaptability to different aperture reflectors, the reflector with a polygonal carbon fiber square-tube back structure is being widely used in HTS[3-5]. Unfortunately, a periodic temperature load change of the reflector in orbit environment will introduce thermal expansion and thermal stress. Surface distortion will be occurred in case of thermal mismatch between reflector components, which is an undesirable situation for the reflector performance[6].

Domestic scholars had studied reflector materials, reflector structures[6-11],and thermal deformation test methods[2,12-13]. We know the deviation of ply angle can cause the non-axisymmetric deformation of the laminated plate, and the deformation will lead to a undesirable thermal deformation of the reflector. Unluckily, we have not carried out relevant work yet. In 1989, foreign scholar Robert[14] proposed a concept of statistical variables to analyze the thermal deformation of high-precision reflectors; William[15] obtained a sample mean value and standard deviation of mechanical property parameters through testing samples, and randomly generated parameter values of each element in the finite element model by the Monte Carlo method. Lang[16] established the displacement deformation gradient matrix based on the linear Gradient Based Superposition method, obtained the relationship between the random variable and the displacement deformation field of the reflector. Meanwhile, increasement of the number of plies could reduce the impact of ply angle deviation on thermal deformation, but it fulfilled at the cost of weight increasement[17].

For the above reasons, we introduce the concept of probability and statistics in this paper, develop the ply angle deviation model of laminated composite materials, and give the thermal deformation analysis method of the reflector based on normal distribution random samples. By using the orthogonal experimental design, the significance of the influence factors of thermal deformation was analyzed, and the better conditions were obtained. According to some random samples subject to normal distribution, the probability level and distribution function of reflector thermal deformation were predicted. This will provide a reference for reflector design and process control as well as offer reference for engineering application.

2. Reflector structure and material

The reflector projection aperture D is Φ2600mm, the offset distance H is 500mm, and the focal length F is 4500mm, as shown in figure 1. The paraboloid equation in the reflector design coordinate system O_m-$X_mY_mZ_m$ is $Z=(X^2+Y^2)/4F$. The reflector is composed of a dish, a polygonal back structure, patches and a deployable arm. The dish is bonded with the back structure through some discretely distributed L-shaped carbon fiber patches, and the deployable arm is connected with the back structure by screws, as shown in figure 2. The reflector is a carbon fiber aluminum honeycomb sandwich structure, composed of front skin, back skin and aluminum honeycomb core. The back structure and the deployable arm are all composed of carbon fiber laminated rectangular square tubes.

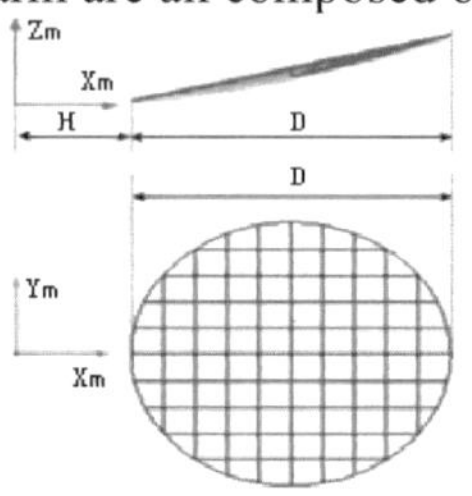

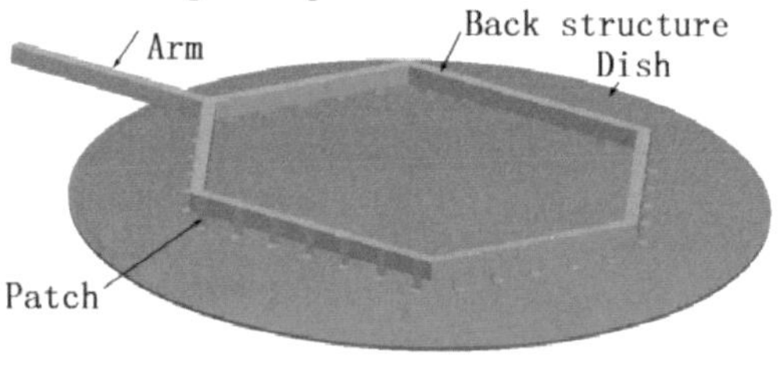

Figure 1. Structure diagram of reflector. **Figure 2.** Model of reflector.

Properties of M40J carbon fiber/epoxy resin prepreg, EC2216A/B adhesive and hexagonal perforated aluminum honeycomb (BC1.8-3/8P-0.038) are shown in table 1.

Table 1. Materials properties of M40J, EC2216A/B and aluminum honeycomb.

Properties		M40J	Glue	Core
Elastic modulus	E_1(GPa)	232	0.90	1
	E_2(GPa)	7.50	/	1
Poisson's ratio	ν_{12}	0.26	0.30	0.3
Shear modulus	G_{12}(GPa)	4.05	0.34	1
	G_{23}(GPa)	/	/	117
	G_{31}(GPa)	/	/	78.9
Coefficient of thermal expansion	$\alpha_1(10^{-6})$	-0.50	100	23.8
	$\alpha_2/(10^{-6})$	41	/	/

3. Thermal deformation analysis model and calculation method

3.1. Finite element model

An analysis model of the reflector is shown in figure 3. The dish, the polygon back structure, the patches and the deployable arm were meshed by Quad4 shell element. The patches was connected with the dish by four MPC/RBE2 units; The deployable arm was connected with the back structure by eight MPC/RBE2 units.

3.2. Load and boundary conditions

The deployable arm is fixed on a two-dimensional pointing mechanism, and the pointing mechanism installed on the satellite deck. Combined with the working conditions of reflector in orbit, we define the constraints as follows: the translational displacement of the four vertices of the deployable arm root segment in the three directions of Xm, Ym and Zm is zero. The temperature load is 120 ℃.

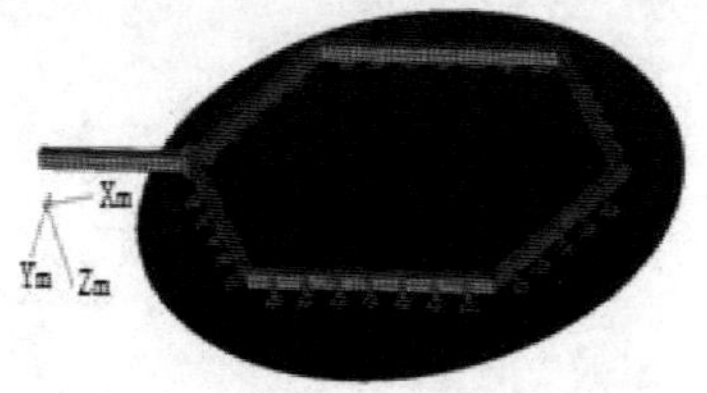

Figure 3. Finite element model of reflector.

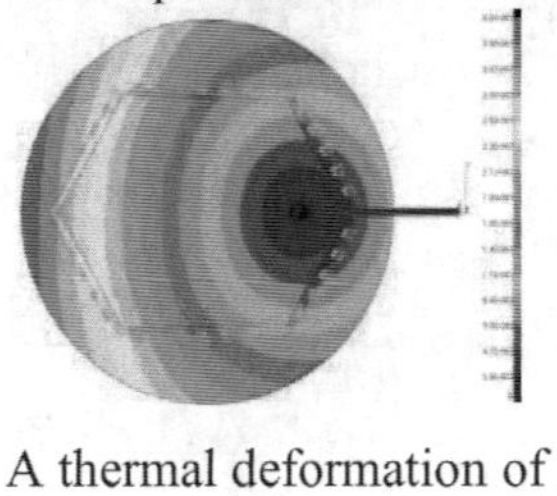

Figure 4. A thermal deformation of reflector.

3.3. Thermal deformation calculation method

A thermal deformation example of polygonal back frame reflector at 120℃is shown in figure 4. Assumed that the number of nodes on the dish is *p*, the initial coordinate is (x_i , y_i , z_i) (*i*=1, 2,..., *p*) at 20℃, and it becomes ($\overline{x}_i$, $\overline{y}_i$, $\overline{z}_i$) at 120℃. To eliminate the rigid displacement caused by the deployable arm, all nodes of the dish after thermal deformation will be done the best fitting than the *p* nodes before thermal deformation, and the fitting value ($\tilde{x}_i$, $\tilde{y}_i$, $\tilde{z}_i$) is found. Best fit variables include rotation angle R_x around the X_m axis and rotation angle R_y around the Y_m axis, because the two-dimensional pointing mechanism has the ability to rotate along the X_m and Y_m axes. So, the root mean square (RMS) value ε of reflector thermal deformation (unit: mm) is shown in equation (1).

$$\varepsilon = \left(\frac{1}{p} \sum_{i=1}^{p} (\tilde{x}_i - x_i)^2 + (\tilde{y}_i - y_i)^2 + (\tilde{z}_i - z_i)^2 \right)^{\frac{1}{2}} \tag{1}$$

3.4. Structure size optimization

The size of back structure is as shown in figure 5, and the value of *L1* and *L2* in size-1 is 1.75m and 1.35m, the value of *L1* and *L2* in size-2 is 1.6m and 1.2m, the value of *L1* and *L2* in size-3 is 1.45m and 1.05m. According to the above result, the influence of the size of the back structure on the thermal deformation of the dish is obtained as shown in figure 6, the ply angles of layer-1 and layer-2 are [0°/±45°/90°/90°/±45°/0°]s and [50°/0°/-50°/0°/0°/-50°/0°/50°]s. It can be seen there is nonlinear stiffness coupling between the back frame and the dish. What's more, compared with the axial zero expansion design, the quasi-isotropic design for the back structure has a better thermostability.

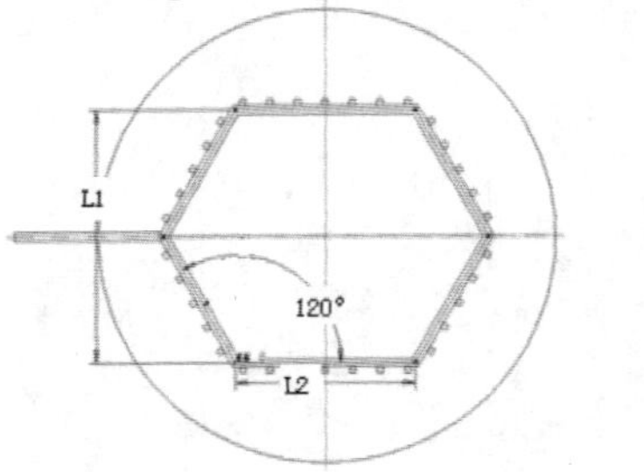

Figure 5. Size of back structure.

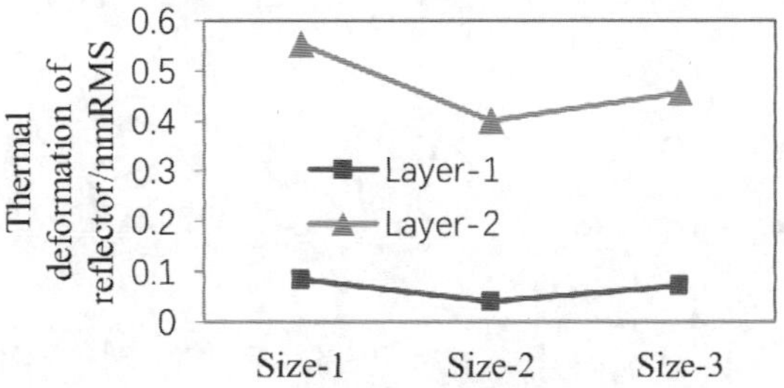

Figure 6. Influence of back structure size on thermal deformation of reflector.

The front skin ply of the dish is [0°/±45°/90°], and the back skin ply is [90°/±45°/0°]. As indicated in figure 7 below, the thermal compatibility with the stiffness of the back structure will be deteriorated with increasing magnitude of heigh of honeycomb core.

4. Thermal deformation calculation model of reflector with ply angle deviation

A configuration of laminated composite structure is shown in figure 8. The total number of layers is n, and the thickness is h. the ply angle of each layer is θ_i (i=1, 2,..., n), the thickness of each layer h_i=h/n. Here, z_0= z_n= h/2.

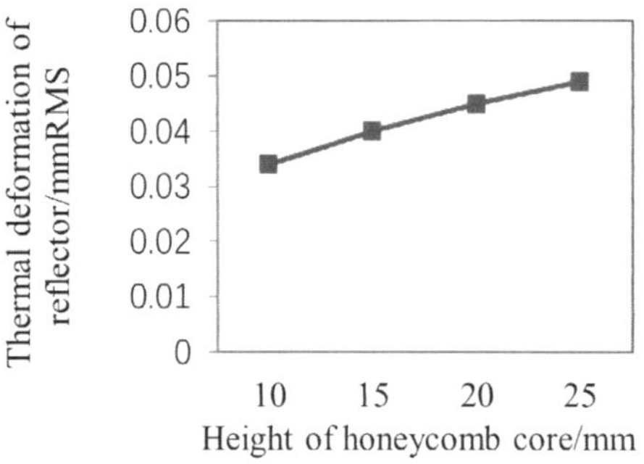

Figure 7. Influence of height of core on thermal deformation of reflector.

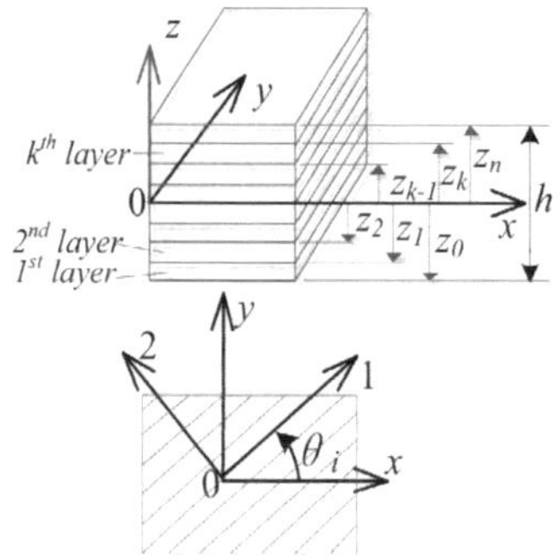

Figure 8. Structural form of laminated composite.

According to the constitutive relationship between generalized internal force and generalized strain of laminated structure[18], it satisfies equation (2).

$$\left(\begin{Bmatrix}\mathbf{N}\\\mathbf{M}\end{Bmatrix}+\begin{Bmatrix}\mathbf{N}^t\\\mathbf{M}^t\end{Bmatrix}\right)=\begin{bmatrix}\mathbf{A}&\mathbf{B}\\\mathbf{B}&\mathbf{D}\end{bmatrix}\begin{Bmatrix}\boldsymbol{\varepsilon}^0\\\boldsymbol{\kappa}\end{Bmatrix} \tag{2}$$

where $\{\mathbf{N}\}$、$\{\mathbf{M}\}$ are the internal force and internal moment per unit width of the laminated structure cross section, respectively; $\{\mathbf{N}^t\}$、$\{\mathbf{M}^t\}$ are the internal force and internal moment caused by temperature change, respectively; When temperature load Δt is applied on the structure, there is $\{\mathbf{N}\}=\{\mathbf{M}\}$=0. Substituting it into equation (2) obtains equation (3). Equations (4)- (5) provide the calculation method for the correlation matrix in Equation (3), there are $l=\cos\theta$ and $m=\sin\theta$.

$$\begin{Bmatrix}\mathbf{N}^t\\\mathbf{M}^t\end{Bmatrix}=\begin{bmatrix}\mathbf{A}&\mathbf{B}\\\mathbf{B}&\mathbf{D}\end{bmatrix}\begin{Bmatrix}\boldsymbol{\varepsilon}^0\\\boldsymbol{\kappa}\end{Bmatrix} \tag{3}$$

$$\begin{cases}\{\mathbf{N}^t\}=\Delta t\sum_{k=1}^{n}\left[\bar{\mathbf{Q}}\right]_k\{\bar{\boldsymbol{\alpha}}\}_k(z_k-z_{k-1})\\\{\mathbf{M}^t\}=\dfrac{\Delta t}{2}\sum_{k=1}^{n}\left[\bar{\mathbf{Q}}\right]_k\{\bar{\boldsymbol{\alpha}}\}_k(z_k^2-z_{k-1}^2)\end{cases} \tag{4}$$

$$\{\bar{\boldsymbol{\alpha}}\}_k=\begin{Bmatrix}\bar{\alpha}_1\\\bar{\alpha}_2\\\bar{\alpha}_{12}\end{Bmatrix}_k=\begin{Bmatrix}\alpha_1l^2+\alpha_2m^2\\\alpha_1m^2+\alpha_2l^2\\2(\alpha_1-\alpha_2)lm\end{Bmatrix} \tag{5}$$

Here, $\begin{bmatrix}\mathbf{A}&\mathbf{B}\\\mathbf{B}&\mathbf{D}\end{bmatrix}$ is the total stiffness matrix. $[\mathbf{A}]$ is the tension-compression stiffness matrix, $[\mathbf{B}]$ is the tension (compression) bending (torsion) coupling stiffness matrix, $[\mathbf{D}]$ is the bending (torsion) stiffness matrix, the calculation method is shown in Equation (6)-(8).

$$\begin{cases} A_{ij} = \sum_{k=1}^{n} (\bar{Q}_{ij})_k (z_k - z_{k-1}) \\ B_{ij} = \frac{1}{2} \sum_{k=1}^{n} (\bar{Q}_{ij})_k (z_k^2 - z_{k-1}^2) (i, j = 1, 2, 6) \\ D_{ij} = \frac{1}{3} \sum_{k=1}^{n} (\bar{Q}_{ij})_k (z_k^3 - z_{k-1}^3) \end{cases} \tag{6}$$

$$\begin{cases} \bar{Q}_{11} = \frac{E_1}{1-\nu_{12}\nu_{21}} l^4 + \frac{E_2}{1-\nu_{12}\nu_{21}} m^4 + \frac{2E_1\nu_{21}}{1-\nu_{12}\nu_{21}} l^2 m^2 + 4G_{12} l^2 m^2 \\ \bar{Q}_{22} = \frac{E_1}{1-\nu_{12}\nu_{21}} m^4 + \frac{E_2}{1-\nu_{12}\nu_{21}} l^4 + \frac{2E_1\nu_{21}}{1-\nu_{12}\nu_{21}} l^2 m^2 + 4G_{12} l^2 m^2 \\ \bar{Q}_{12} = \frac{E_1 l^2}{1-\nu_{12}\nu_{21}} (m^2 + \nu_{21} l^2) + \frac{E_2 m^2}{1-\nu_{12}\nu_{21}} (l^2 + \nu_{12} m^2) + \frac{2E_1\nu_{21}}{1-\nu_{12}\nu_{21}} l^2 m^2 - 4G_{12} l^2 m^2 \\ \bar{Q}_{16} = \frac{E_1(1-\nu_{21})}{1-\nu_{12}\nu_{21}} l^3 m + \frac{E_2(1-\nu_{21})}{1-\nu_{12}\nu_{21}} l m^3 - 2G_{12}(l^2 - m^2) l m \\ \bar{Q}_{26} = \frac{E_1(1-\nu_{21})}{1-\nu_{12}\nu_{21}} l m^3 + \frac{E_2(1-\nu_{21})}{1-\nu_{12}\nu_{21}} l^3 m + 2G_{12}(l^2 - m^2) l m \\ \bar{Q}_{66} = \frac{E_1(1-\nu_{21})}{1-\nu_{12}\nu_{21}} l^2 m^2 + \frac{E_2(1-\nu_{21})}{1-\nu_{12}\nu_{21}} l^2 m^2 + G_{12}(l^2 - m^2) \\ \nu_{21} = E_2 \nu_{12} / E_1 \end{cases} \tag{7}$$

$$\begin{cases} l^4 = \frac{1}{8}(3 + 4\cos 2(\theta_i + \Delta\theta_i) + \cos 4(\theta_i + \Delta\theta_i)) \\ l^2 m^2 = \frac{1}{8}(1 - \cos 4(\theta_i + \Delta\theta_i)) \\ m^4 = \frac{1}{8}(3 - 4\cos 2(\theta_i + \Delta\theta_i) + \cos 4(\theta_i + \Delta\theta_i)) \end{cases} \tag{8}$$

Take one sample of [1°/46°/-45°/89°/C15/90°/-44°/46°/-1°] (C15 represents the height of aluminum honeycomb core) as an example, substitute the values in Equations (9), (10) and table 2 into the finite element model and assign the value of the dish pavement.

Table 2. Change of coefficient of thermal expansion caused by angle deviation.

CTE	Theoretical ply angle	ply angle with deviation
$\bar{\alpha}_{1c}$ (10^{-6})	1.11	1.13
$\bar{\alpha}_{2c}$ (10^{-6})	1.11	1.09
$\bar{\alpha}_{12c}$ (10^{-6})	0.00	-0.01

According to equation (1), the RMS value of reflector thermal deformation are 0.027mm and 0.135mm, respectively. Obviously, it can be seen that there are the following changes after considering the ply angle deviation.

- $[\mathbf{B}] \neq 0; A_{11} \neq A_{22}, A_{66} \neq (A_{11} - A_{12})/2$;
- The value of matrix $[\mathbf{D}]$ changes;
- The deterioration of thermal deformation caused by the ply angle deviation cannot be ignored.

$$\begin{bmatrix} \mathbf{A} & \mathbf{B} \\ \mathbf{B} & \mathbf{D} \end{bmatrix} = \begin{bmatrix} 148.0 & 47.1 & 0 & 0 & 0 & 0 \\ 47.1 & 148.0 & 0 & 0 & 0 & 0 \\ 0 & 0 & 50.5 & 0 & 0 & 0 \\ 0 & 0 & 0 & 37.4 & 9.6 & 0.9 \\ 0 & 0 & 0 & 9.6 & 26.6 & 0.9 \\ 0 & 0 & 0 & 0.9 & 0.9 & 10.3 \end{bmatrix} \tag{9}$$

$$\begin{bmatrix} \mathbf{A} & \mathbf{B} \\ \mathbf{B} & \mathbf{D} \end{bmatrix} = \begin{bmatrix} 147.2 & 47.1 & \approx 0 & 0.1 & \approx 0 & \approx 0 \\ 47.1 & 148.8 & 0.2 & \approx 0 & -0.1 & -0.6 \\ \approx 0 & 0.2 & 50.5 & \approx 0 & -0.6 & \approx 0 \\ 0.1 & \approx 0 & \approx 0 & 37.2 & 9.6 & 1.1 \\ \approx 0 & -0.1 & -0.6 & 9.6 & 26.8 & 1.0 \\ \approx 0 & -0.6 & \approx 0 & 1.1 & 1.0 & 10.3 \end{bmatrix} \tag{10}$$

5. Thermal deformation analysis based on normal distribution random samples

Firstly, there is no thermal deformation coupling effect between the deployable arm and the back structure. Secondly, the structure size of patches is very small, and the influence of the discreteness of the ply angle deviation is relatively small. Therefore, when doing the thermal deformation analysis, the angle deviation between the dish and the back structure is mainly discussed. And it is assumed that:

1) The ply angle deviation of dish and back structure is subject to the normal distribution model, and each ply angle is independent of each other; The theoretical ply of dish is [0°/45°/-45°/90°/C15/90°/-45°/45°/0°]. Meanwhile, the theoretical ply of six tubes of the back structure is [0°/45°/-45°/90°/90°/-45°/45°/45°/0°]s. Then, the random deviation of each ply angle of reflector is $\triangle\theta_i$ (i=1, 2, ..., 104) and satisfies equation (11).

$$f(\Delta\theta_i)=\frac{1}{\sqrt{2\pi}\sigma}\exp(-\frac{(\Delta\theta_i-\mu)^2}{2\sigma^2}) \tag{11}$$

2) According to the normal distribution and the required value of ply angle deviation $\pm\delta$ (δ=0.5°, 1°), it can be generated N sample values of deviation angle with M layers by equation (12) in @Matlab software. In terms of the μ test method[19], take N equal to 100.

$$\Delta\theta_i = \mathrm{randi}\left(\left[-\delta \quad +\delta\right], N, M\right) \tag{12}$$

3) On the basis of random samples of ply angle deviation, update each ply angle according to equation (13), analyze thermal deformation and obtain thermal deformation results ε_i (i=1, 2, ..., N).

$$\theta_i = \theta_i + \Delta\theta_i \tag{13}$$

4) Based on the maximum likelihood estimation method, the mean value $\hat{\mu}_\varepsilon$ and standard deviation $\hat{\sigma}_\varepsilon$ of the reflector thermal deformation samples are shown in equations (14) and (15).

$$\hat{\mu}_\varepsilon=\frac{1}{N}\sum_{i=1}^{N}\varepsilon_i=\overline{\varepsilon} \tag{14}$$

$$\hat{\sigma}_\varepsilon=\frac{1}{N}\sum_{i=1}^{N}(\varepsilon_i-\overline{\varepsilon})^2=s_N^2 \tag{15}$$

5.1. Othogonal experimental design

According to the orthogonal test design method[19], selected an orthogonal table L9(3^4), the orthogonal test results and the analysis of variance were shown in table 3 and table 4, respectively. The horizontal number factors contained A (ply angle deviation of dish) and B (ply angle deviation of back structure), where level 1 represents δ=0°. Level 2 represents δ = 0.5°, and level 3 represents δ = 1°. $A \times B$ is the interaction between factors A and B. R is the range, s_i^2 is the sum of squares of changes in the column, f_i is the degree of freedom in the column, e is the error, and T is the sum of all data.

Table 3. Results of orthogonal test.

Row / *Test*	*A* 1	*B* 2	*A×B* 3	4	ε_i ×10³
1	1	1	1	1	27
2	1	2	2	2	121
3	1	3	3	3	25
4	2	1	2	3	54
5	2	2	3	1	81
6	2	3	1	2	52
7	3	1	3	2	136
8	3	2	1	3	46
9	3	3	2	1	135
I	173	217	125	243	T=677
II	187	248	310	309	
III	317	212	242	125	
R	144	36	205	84	
s_i^2	4204	254	5838	5792	

Table 4. Analysis of variance.

Factor	s_i^2	f_i	s_i^2 / f_i	*F*	*Significant Difference*
A	4204	2	2102	0.73	*
B	254	2	177	0.06	
A×B	5838	2	2919	1.01	**
e	5792	2	2896		

The analysis results are as follows:

a) An interaction on the thermal deformation is found between the dish and the back structure, that means the back structure has a coupling effect on the thermal deformation of the dish.

b) The interaction on the thermal deformation is the most significant, the dish takes the second place, and the significance of the back structure is the weakest.

c) Because $A \times B$ interaction is significant, consider unbiased statistics in equation (16), so $(\hat{a}b)_{11}=(\hat{a}b)_{23}=(\hat{a}b)_{32}$ is better. In fact, the angle deviation of paving layer is not zero, and factor A is significant, so the better case is A_2B_3.

$$(\hat{a}b)_{11}=(\hat{a}b)_{23}=(\hat{a}b)_{32}<(\hat{a}b)_{13}=(\hat{a}b)_{22}=(\hat{a}b)_{31}<(\hat{a}b)_{12}=(\hat{a}b)_{21}=(\hat{a}b)_{33} \quad (16)$$

5.2. Analysis and discussion

Based on the above result, simulation results of reflector thermal deformation are shown in figure 9(a), RMS value of thermal deformation is distributed in the range of [0.038, 0.119]. The columnar statistical results are shown in figure 9(b), $\hat{\mu}_\varepsilon$ is 0.069mm, $\hat{\sigma}_\varepsilon$ is 0.014mm.

Results are shown in table 5 according to χ2-test of goodness of fit. Because χ_N^2=11.02< $\chi_{0.05}^2(6)$ =12.6, so the thermal deformation distribution of the reflector is normal distribution, as shown in equation (17).

$$f(\varepsilon)=\frac{1}{\sqrt{2\pi}\times 0.014}\exp\left(-\frac{(\varepsilon-0.069)^2}{2\times 0.014^2}\right) \quad (17)$$

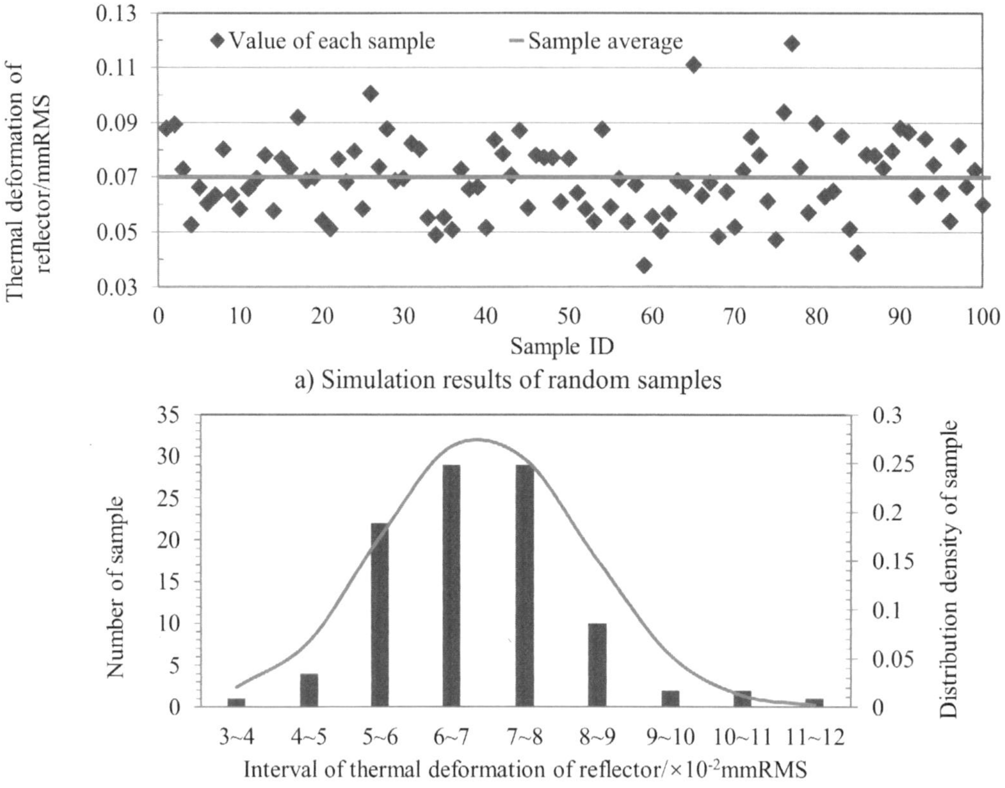

a) Simulation results of random samples

b) Histogram for sample RMS results

Figure 9. Reflector thermal deformation analysis results.

Table 5. χ2-test of goodness of fit.

Interval ε_i	FRQ N_i	Normalized interval $\left[\mu_{\varepsilon_i}, \mu_{\varepsilon_{i+1}}\right]$	$\hat{p}_{i0}$	$N\hat{p}_{i0}$	$(N_i\text{-}N\hat{p}_{i0})$	$(N_i\text{-}N\hat{p}_{i0})^2$	$\frac{(N_i\text{-}N\hat{p}_{i0})^2}{N\hat{p}_{i0}}$
0.03~0.04	1	(-∞, -2.07)	0.020	2.0	-1	1	0.50
0.04~0.05	4	[-2.07, -1.36)	0.067	6.7	-2.7	7.29	1.09
0.05~0.06	22	[-1.36, -0.64)	0.174	17.4	4.6	21.16	1.22
0.06~0.07	29	[-0.64, +0.07)	0.267	26.7	2.3	5.29	0.20
0.07~0.08	29	[+0.07, +0.79)	0.254	25.4	3.6	12.96	0.51
0.08~0.09	10	[+0.79, +1.50)	0.151	15.1	-5.1	26.01	1.72
0.09~0.10	2	[+1.50, +2.21)	0.053	5.3	-3.3	10.89	2.05
0.10~0.11	2	[+2.21, +2.93)	0.012	1.2	0.8	0.64	0.53
0.11~0.12	1	[+2.93, +∞)	0.002	0.2	0.8	0.64	3.20
Σ			1.000	100			11.02

5.3. Verification and discussion

According to the corresponding material, ply design and ply angle deviation control, the 2.6m polygonal back structure reflector was manufactured. The thermal deformation test was carried out using the photogrammetric method[12-13], and the equipment was the MPS/FP single-camera industrial photogrammetric system. Test status in the thermal vacuum tank is shown in figure 10.

1) The test result is 0.075mmRMS according to equation (1), and is distributed in the sample interval [0.07, 0.08). The displacement deformation is shown in figure 11. The relative error is only 8.7% between the simulation result with a value 0.069mmRMS and the test.

Figure 10. Thermal deformation test status.

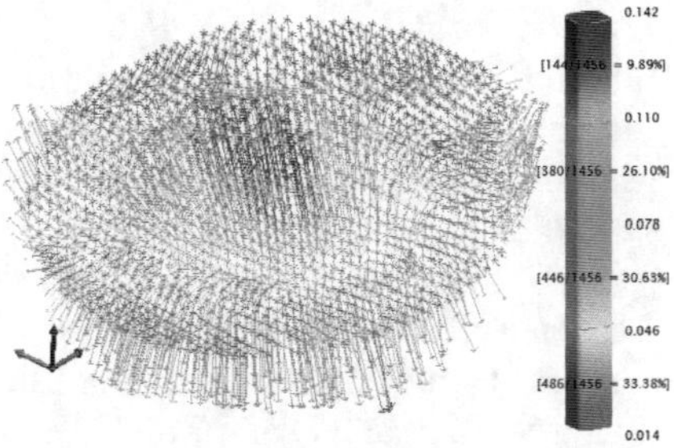

Figure 11. Thermal deformation test result.

2) According to the same method in this paper, the thermal deformation analysis result of the dish is only 0.061mm RMS based on the ±0.5° ply angle deviation. Compared with 0.069mmRMS, the thermal coupling effect between the dish and the back structure is only 13.1% of the thermal deformation of the dish, as indicated in figure 12. It shows that the thermal coupling effect between the back structure and the dish can be improved by optimizing the ply angle deviation. At the same time, with the increase of the ply-angle deviation, the nonlinear of the dish thermal deformation value greatly increases.

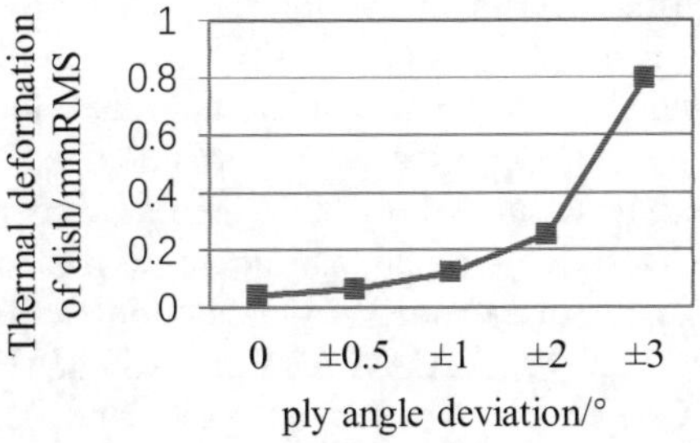

Figure 12. Influence of ply angle deviation on thermal deformation of dish

6. Conclusion

In this paper, based on the ply angle deviation model and the idea of probability statistics, a method of reflector thermal deformation analysis is proposed. Because the ply angle deviation is difficult to obtain in the actual manufacturing process, so random samples subject to normal distribution is given as well as used as simulation samples. Meanwhile, $\chi 2$-test of goodness of fit is applied to verify the correctness of the model. Results show that simulation values are consistent with experimental values. So it can be applied for engineering applications in the future.

References

[1] H Fenech, S Amos and A Hirsch. VHTS systems: requirements and evolution [C], The 11th European Conference on Antennas and Propagation, Paris, France, March 19-24, 2017.

[2] Jiang W J, Shi J W and Yu X M. Manufacture of large aperture and high precision composite reflector[J]. Journal of Nanjing University of Aeronautics & Astronautics, 2019, 51(S): pp 22-26.

[3] Nicolas E and Valérie D. Industrial antenna reflector at EADS Space[C]. The 1th European Conference on Antennas and Propagation, Nice, France, Nov. 6-10, 2006.

[4] N Elie, L Ravelli and A. Lacombe， et a1. Antenna reflectors for telecommunication satellites at EADS Astrium Space transportation LES MUREAUX[C]. The 2nd European Conference on Antennas and Propagation, Edinburgh, UK, NOV. 11-16, 2007.

[5] Duval H and Leborgne F. 5 meter truncated solid shell reflector: Arianegroup alternative to expensive deployable mesh reflectors[C]. The 70th International Astronautical Congress, Washington, DC, USA, Oct 11-16, 2019.

[6] Fang H F, Huang P M and Zhou Y, et al. Analytical investigation of a high precision reflector[C]. Spacecraft Structures Conference. National Harbor, Maryland, Jan.13-17, 2014.

[7] Zhou T, Ye Z J and Shi Y H, et al. Thermal distortion for honeycomb sandwich construction rigid reflector of satellite antenna[J]. Acta Materiae Compositae Sinica, 2018, 35(8): pp 2065-73.
[8] Zhang H C, Liang X C and Wang X L, et al. Simulation analysis on the profile accuracy of parabolic antenna with honeycomb sandwich structure [J]. Aerospace Shanghai, 2020, 37(1): pp 107-112.
[9] Zhang C, Wei J F and Ma X F, et al. Optimization of thermal deformation of ultrahigh frequency remote sensing reflector and manufacture of a zero-expansion composite[J]. Acta Materiae Compositae Sinica, 2019, 36(4): pp 795-800.
[10] Zhang W B, Zhang Q and Xu H T, et al. Structure design of carbon fiber grid sandwich reflector with high stability and optimization of thermal deformation[J]. Composites Science and Engineering, 2020, (5): pp 40-46.
[11] Zhang Z Y, Jiang S C and Wang M, et al. Analysis of influencing factors on thermal deformation of satellite onboard antenna reflector[J]. Spacecraft Environment Engineering, 2021, 38(2): pp 130-137.
[12] Wan X P and Yu X Z. Test and analysis of thermal distortion of a deployable CFRP reflector[J]. Spacecraft Environment Engineering, 2016, 33(6): pp 672-675.
[13] Ma Z X, Yang D H and Wang S Q, et al. Antenna reflector surface fitting algorithm based on rigid body displacement principle[J]. Chinese Journal of Mechanical Engineering, 2010, 46(18): pp 29-35.
[14] Robert E F, Paul M M and Robert D J. Technical approach for the development of structural composite mirrors[C]. Active Telescope Systems. Orlando, FL, Sep.20, 1989.
[15] William W W. A probabilistic model of the thermal dimensional stability of composite structural components[C]. Design of optical instruments. Orlando, FL, Apr. 22-24, 1992.
[16] Lang M, Baier H and Ernst T. Surface accuracy prediction and optimization of a high precision Q/V-Band reflector [C]. The 47th AIAA/ASME /ASCE/AHS/ASC Structures, Structural Dynamics, and Materials Conference, Newport, Rhode Island, May 1-4, 2006.
[17] Kuo C P, Lou M C and Rapp D. Thermal-mechanical behavior of high precision composite mirrors[J], AIAA, 1993.(1411): pp 949-958.
[18] Chen L M and Yang B N. Mechanical analysis of composite materials[M]. Beijing: China Science and Technology Press, 2006.
[19] Shi Y M, Xu W and Qin C Y, et al. Mathematical statistics (Third Edition)[M]. Beijing: Science Press, 2009.

Multiple fault isolation method for micro thrusters of drag-free systems

Haoran Li[1,2], Xiaodong Shao[3,*], Qinglei Hu[1,2], Dong Zhao[4], Yonghe Zhang[5], Pengcheng Wang[5], Zhaohui Dang[6] and Dongyu Li[1,4]

[1] Tianmushan Laboratory, Hangzhou 310023, China
[2] School of Automation Science and Electrical Engineering, Beihang University, Beijing 100191, China
[3] School of Aeronautic Science and Engineering, Beihang University, Beijing 100191, China
[4] School of Cyber Science and Technology, Beihang University, Beijing 100191, China
[5] Innovation academy for microsatellites of Chinese Academy of Sciences, Shanghai 201203, China
[6] School of Astronautics, Northwestern Polytechnical University, Xi'an 710072, China

* Corresponding author's email: xdshao_sasee@buaa.edu.cn

Abstract. This paper proposes a multiple fault isolation framework for micro thrusters of drag-free systems, based on the minimum number of isolation filters. Based on the consideration of fault isolatability, the original system is reasonably divided into two completely isolated subsystems, and then two sets of fault isolation filters are designed based on the principles of duality design for state feedback decoupling, which achieves residual decoupling, and therefore, can detect and isolate multiple faults occurred on the micro thrusters. In particular, this method has low computational burden (a feature of practical importance in reducing the usage of onboard resources) and short isolation time, making it suitable for on-board fast and precise fault isolation. The simulation results demonstrate the efficacy of the proposed method when utilized for a drag-free system with 12 field emission electric propulsion (FEEP) thrusters.

1. Introduction

The space gravitational wave detection system is a multi-body system that achieves stable formation flight through a million-kilometer separation between three spacecrafts, and it carries six test masses (TMs) that require nano-precision motion control, which needs to be provided by drag-free control technique. Figure 1 shows the drag-free system of LISA Pathfinder (LPF). In recent decades, various countries have conducted some studies on the drag-free control problem in space gravitational wave detection, which can be found in [1-3]. The field emission electric propulsion (FEEP) thrusters are the core of the drag-free satellite. In space gravitational wave detection task, the spacecraft is equipped with 12 FEEP thrusters to generate μN magnitude thrust to achieve ultra-high precision drag-free control.

According to the task requirements, the drag-free satellite need to operate on orbit to wait for the space gravitational wave detection window, which makes it susceptible to faults especially the thrusters because of the harsh environment of the space. Thus, it is of significance to conduct fault diagnosis for the thrusters of the drag-free system. In the past decades, many studies have been conducted on fault diagnosis, which could be divided into three categories: model-based methods, signal processing methods, and artificial intelligence methods, in which model-based methods are the most suitable for the satellite on-board fault diagnosis because of its reliability and computation burden. Much work are devoted to study fault diagnosis problem about the satellite system. Tudoroiu and Khorasani [4] designed a fault diagnosis algorithm using a bank of Kalman filters for the satellite attitude control system's actuators. Henry [5] considered satellite thruster faults and designed two kinds of robust fault diagnosis algorithms based on Linear Matrix Inequality (LMI) techniques.

Figure 1. LPF drag-free system. Image courtesy: ESA.

However, most research is focused on 6 degrees-of-freedom (DOF) satellites, with limited studies on fault diagnosis for the drag-free system. Fault diagnosis for thrusters of the drag-free system has two difficulties: 1) the drag-free system has a complex dynamics model with high DOF; 2) the system carries many thrusters which are not paired arranged, thus it is difficult to isolate all thrusters faults meanwhile. The work on fault isolation mainly include three kinds of methods [6]: 1) unknown input decoupling methods; 2) the dual problem solution of decoupled controller design; 3) designing a bank of residuals generators. Among them, the first methods mainly treat the faults in the remaining dimensions and all disturbances as unknown inputs. It then decouples them from the current dimension's fault signal through unknown input observers (UIO). To achieve this, it is necessary for the number of measured system outputs to be greater than or equal to the dimensions of the unknown inputs [7]. However, since the dimension of the fault signal caused by the thrusters in the drag-free system is relatively high, this method is not suitable for fault diagnosis in such systems. The third methods need to design a bank of 12 observers [8]. Due to the high dimensionality of the observer, running it in parallel would result in significant computational resource consumption.

The second methods consider the fault isolation problem as the dual problem of state feedback decoupling design. By designing a set of observer gain matrix and post-filter matrix, the fault signals could achieve mutual decoupling. This method was originally proposed by Liu and Si in [9]. This method only requires the design of one observer when applied, leading to small computational resource consumption. The algorithm is simple, efficient, and highly reliable. However, it has a finite maximum number of faults that can be isolated, and it needs to satisfy certain rank conditions. Li and Jaimoukha [10] consider the robust fault isolation problem against the disturbances using H_∞ norm based on [9], and applied to a 4-order dynamic system. However, this method used LMI in the solving process, which can pose difficulties in solving problems for high-dimensional drag-free systems.

Inspired by the aforementioned research, this paper proposes a fault isolation framework which is suitable for micro thrusters of the drag-free system. This framework can detect and isolate multiple faults occurred on the 12 FEEP thrusters using the minimum number of filters, which saves computational resources. First, a 15 DOF dynamic model of the drag-free system is established considering the thrusters faults, and model-based fault isolation problem of the drag-free system is formulated. Next, a control algorithm is applied to the drag-free satellite to obtain a close-loop system. And considering the fault isolability of drag-free systems, the original system is divided into as few subsystems as possible, with each subsystem being fully fault-isolatable. Based on the design method of duality of state feedback decoupling, a bank of parallel fault isolation filters towards all the subsystems is designed to isolate all the thrusters faults. It is worth noting that the number of filters is minimum and much less than 12, which can greatly save onboard resources.

The remainder of this paper is organized as follows. Section 2 presents the dynamics of the LISA Pathfinder (LPF) drag-free satellite, which includes the thrusters configuration and fault models. In Section 3, we introduce a fault isolation framework based on parallel observer. Simulation results are shown in Section 4, followed by the conclusions and future works.

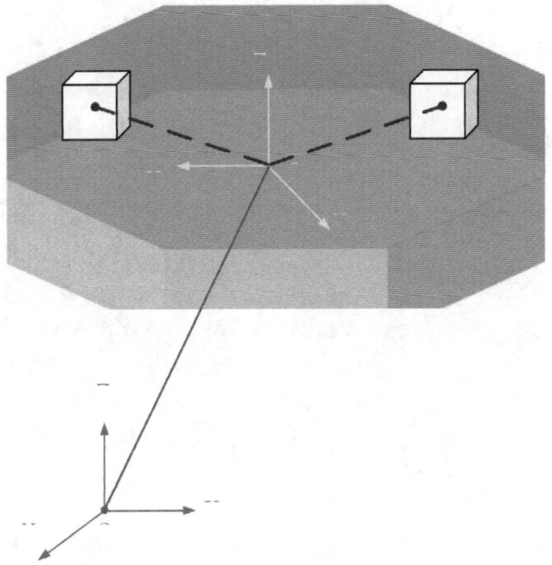

Figure 2. Coordinate frame systems.

2. Drag-free system dynamics

In this section, a 15-DOF dynamic model of the drag-free system is established. Then, the thruster configuration is presented with explicit consideration of four types of thruster faults. Finally, the fault isolation problem of the micro thrusters of the drag-free system is formulated as an observer design problem based on fault detection filter (FDF).

2.1. Dynamics of drag-free systems

We take the LPF drag-free system as a case study, which a three-body system consisting of a spacecraft (S/C) and two test masses (TMs), as shown in figure 2. Before proceeding, we define two coordinate systems: the heliocentric inertial (HCI) frame and the body-fixed (BF) frame of the satellite.

1) Heliocentric inertial frame $\mathcal{I}$. Since the LPF satellite operates in the heliocentric orbit, we chose the HCI frame as the inertial reference frame. The center of the Sun serves as the origin $O_{\mathcal{I}}$, and the $\boldsymbol{X}_{\mathcal{I}}$ axis is aligned with the ecliptic plane and points towards the J2000 equinox. The $\boldsymbol{Z}_{\mathcal{I}}$ axis is perpendicular to the eclipticplane, while the $\boldsymbol{Y}_{\mathcal{I}}$ axis completes the right-handed coordinate frame.

2) Spacecraft body-fixed frame $\mathcal{S}$. The origin O_{S} of the coordinate system is fixed on the center of mass (CoM) of the satellite, aligning the triad axes with its principal axes of inertia. The $\boldsymbol{X}_{S}$ axis is parallel to the line between the test mass 1 (TM1) and the test mass 2 (TM2), and points from TM2 to

TM1. The $\boldsymbol{Z}_S$ axis is perpendicular to the upper plane of the satellite, and the $\boldsymbol{Y}_S$ axis completes the right-handed coordinate frame.

Based on the assumptions of rigid body and small-angle maneuvers, the 15-DOF dynamics of the LPF satellite is modeled as [11]

$$\begin{bmatrix} \ddot{\boldsymbol{\varphi}} \\ \boldsymbol{r}_1 \\ \boldsymbol{r}_2 \\ \ddot{\boldsymbol{\varphi}}_1 \\ \ddot{\boldsymbol{\varphi}}_2 \end{bmatrix} = \begin{bmatrix} \mathbf{0} & \boldsymbol{E} & \mathbf{0} & \mathbf{0} & \mathbf{0} & \mathbf{0} \\ -\boldsymbol{E} & \tilde{\boldsymbol{r}}_{O1} & \boldsymbol{E} & \mathbf{0} & \mathbf{0} & \mathbf{0} \\ -\boldsymbol{E} & \tilde{\boldsymbol{r}}_{O2} & \mathbf{0} & \mathbf{0} & \boldsymbol{E} & \mathbf{0} \\ \mathbf{0} & -\boldsymbol{E} & \mathbf{0} & \boldsymbol{E} & \mathbf{0} & \mathbf{0} \\ \mathbf{0} & -\boldsymbol{E} & \mathbf{0} & \mathbf{0} & \mathbf{0} & \boldsymbol{E} \end{bmatrix} \begin{bmatrix} \boldsymbol{a} \\ \boldsymbol{\alpha} \\ \boldsymbol{a}_1 \\ \boldsymbol{\alpha}_1 \\ \boldsymbol{a}_2 \\ \boldsymbol{\alpha}_2 \end{bmatrix} \tag{1}$$

where $\boldsymbol{\varphi} = [\phi, \theta, \eta]^\top$ is the attitude of the S/C, whereas as $\boldsymbol{\varphi}_i = [\phi_i, \theta_i, \eta_i]^\top$ and $\boldsymbol{r}_i = [x_i, y_i, z_i]^\top$ $(i = 1,2)$ are the relative attitude and position of the ith test mass (TM) with respect to their nominal attitude and position in their housing in frame $\mathcal{S}$. Here, $\tilde{\boldsymbol{r}}_{Oi}$ represents the skew-symmetric matrix of $\boldsymbol{r}_{Oi}$ in the frame $\mathcal{S}$, that is

$$\tilde{\boldsymbol{r}}_{Oi} = \begin{bmatrix} 0 & -r_{Oi,z} & -r_{Oi,y} \\ r_{Oi,z} & 0 & -r_{Oi,x} \\ r_{Oi,y} & r_{Oi,x} & 0 \end{bmatrix} \tag{2}$$

The matrix $\boldsymbol{E}$ is the unit matrix of appropriate dimension, $\boldsymbol{a}$, $\boldsymbol{\alpha}$, $\boldsymbol{a}_i$, $\boldsymbol{\alpha}_i$ denote the accelerations of spacecraft and the ith TM. They can be further defined in a more concise form, i.e.

$$\begin{aligned} \boldsymbol{a}_{SC} &= [\boldsymbol{a}, \boldsymbol{\alpha}]^\top = \boldsymbol{M}^{-1}[\boldsymbol{f}, \boldsymbol{l}]^\top \\ \boldsymbol{a}_{TM1} &= [\boldsymbol{a}_1, \boldsymbol{\alpha}_1]^\top = \boldsymbol{M}_1^{-1}[\boldsymbol{f}_1, \boldsymbol{l}_1]^\top \\ \boldsymbol{a}_{TM2} &= [\boldsymbol{a}_2, \boldsymbol{\alpha}_2]^\top = \boldsymbol{M}_2^{-1}[\boldsymbol{f}_2, \boldsymbol{l}_2]^\top \end{aligned} \tag{3}$$

where $\boldsymbol{M}$ and $\boldsymbol{M}_i\,(i=1,2)$ are the mass matrices of the S/C and the two TMs, $\boldsymbol{f}$ and $\boldsymbol{f}_i\,(i=1,2)$, $\boldsymbol{l}$ and $\boldsymbol{l}_i\,(i=1,2)$ are the total force and torque applying on the S/C and TMs, which can be further divided into two parts: actuation and stiffness contributions

$$\begin{aligned} \boldsymbol{a}_{SC} &= \boldsymbol{u}_{SC} \\ \boldsymbol{a}_{TM1} &= \boldsymbol{u}_{ES1} - \boldsymbol{\Phi}_{DF}^2 \boldsymbol{q}_{DF} \\ \boldsymbol{a}_{TM2} &= \boldsymbol{u}_{ES2} - \boldsymbol{\Phi}_{SUS}^2 \boldsymbol{q}_{SUS} \end{aligned} \tag{4}$$

where $\boldsymbol{\Phi}_j^2, j = DF, SUS$ are the stiffness matrices. According to the task requirements, drag-free and suspension state vectors can be defined as

$$\begin{aligned} \boldsymbol{q}_{DF} &:= [x_1, y_1, z_1, \phi_1, y_2, z_2]^\top \\ \boldsymbol{q}_{SUS} &:= [\theta_1, \eta_1, x_2, \phi_2, \theta_2, \eta_2]^\top \end{aligned} \tag{5}$$

Thus, the dynamic model of the drag-free system can be expressed as follows

$$\begin{bmatrix} \ddot{\boldsymbol{\varphi}} \\ \ddot{\boldsymbol{q}}_{DF} \\ \ddot{\boldsymbol{q}}_{SUS} \end{bmatrix} = \begin{bmatrix} \boldsymbol{B}_{ATT} & \mathbf{0} & \mathbf{0} \\ \boldsymbol{B}_{DF} & \boldsymbol{E} & \mathbf{0} \\ \boldsymbol{B}_{SUS} & \mathbf{0} & \boldsymbol{E} \end{bmatrix} \begin{bmatrix} \boldsymbol{a}_{SC} \\ \boldsymbol{a}_{TM1} \\ \boldsymbol{a}_{TM2} \end{bmatrix} \tag{6}$$

where $\boldsymbol{B}_{ATT}$, $\boldsymbol{B}_{DF}$ and $\boldsymbol{B}_{SUS}$ are appropriate matrices associated with the S/C and TMs, which can be derived by equations (1)-(5).

2.2. Thruster configuration and fault modeling

The drag-free system is outfitted with 12 FEEP thrusters, organized into three clusters of four thrusters each, to achieve ultra-high precision control. This study solely considers additive faults occuring on the 12 FEEP thrusters. The various thruster faults typically fall into four categories, as classified by [12]:
1) Stuck-open: the thruster consistently delivers maximum thrust, regardless of the required demand;
2) Blocked-closed: the thruster does not generate any thrust, regardless of the level of demand;
3) Loss: the thruster encounters a decrease in efficiency;
4) Bias: the actual thrust produced by the thruster deviates slightly from the commanded value.

An additive approach can be utilized to model the above faults

$$\boldsymbol{u}^f(t) = \boldsymbol{u}(t) - \boldsymbol{\zeta}(t) + \boldsymbol{u}_b(t) \tag{7}$$

where $\boldsymbol{u}^f(t)$ represents the actual output of the thrusters, $\boldsymbol{u}(t)$ denotes the commanded control inputs, $\boldsymbol{u}_b(t)$ is the bias-type fault, $\boldsymbol{\zeta}(t) = [\zeta_1(t), \zeta_2(t), \ldots, \zeta_{12}(t)]^\top$ is the fault indicator associated with different fault types, i.e.

$$\zeta_i(t) = \begin{cases} u_{\max} - u_i(t), & \text{if stuck-open} \\ \varepsilon_i u_i(t), & \text{if loss} \\ u_i(t), & \text{if blocked-closed} \\ 0, & \text{if fault free} \end{cases} \tag{8}$$

where ε_i is the loss rate of the ith thruster, $u_{\max}$ is the maximum output provided by thrusters.

According to equations (6)-(8), a linear time-invariant (LTI) state space model of the drag-free system can be derived as follows

$$\begin{aligned} \dot{\boldsymbol{x}} &= \boldsymbol{A}\boldsymbol{x} + \boldsymbol{B}\boldsymbol{u} + \boldsymbol{B}_f \boldsymbol{f} \\ \boldsymbol{y} &= \boldsymbol{C}\boldsymbol{x} \end{aligned} \tag{9}$$

where $\boldsymbol{x} \in \mathbb{R}^{30\times1}$ is the state vector, $\boldsymbol{u} = [\boldsymbol{u}_{DF}, \boldsymbol{u}_T, \boldsymbol{u}_{SUS}]^\top \in \mathbb{R}^{24\times1}$ is control input, $\boldsymbol{f} \in \mathbb{R}^{12\times1}$ is the thruster fault vector. $\boldsymbol{A}$, $\boldsymbol{B}$, $\boldsymbol{C}$ and $\boldsymbol{B}_f$ are system matrices of appropriate dimensions.

2.3. Fault isolation problem formulation

Based on the FDF technique and the state space model (9), we design a state observer to generate fault isolation residuals

$$\begin{aligned} \dot{\hat{\boldsymbol{x}}} &= \boldsymbol{A}\hat{\boldsymbol{x}} + \boldsymbol{B}\boldsymbol{u} + \boldsymbol{R}(\boldsymbol{y} - \boldsymbol{C}\hat{\boldsymbol{x}}) \\ \hat{\boldsymbol{y}} &= \boldsymbol{C}\hat{\boldsymbol{x}} \end{aligned} \tag{10}$$

The residual for fault isolation is defined as

$$\boldsymbol{r} := \boldsymbol{H}(\boldsymbol{y} - \hat{\boldsymbol{y}}) \tag{11}$$

Let us define $\boldsymbol{\xi}(t) := \boldsymbol{x}(t) - \hat{\boldsymbol{x}}(t)$ as the state estimation error. In view of (9)-(11), the system residuals are expressed as

$$\begin{aligned} \dot{\boldsymbol{\xi}}(t) &= (\boldsymbol{A} - \boldsymbol{R}\boldsymbol{C})\boldsymbol{\xi}(t) + \boldsymbol{B}_f \boldsymbol{f}(t) \\ \boldsymbol{r}(t) &= \boldsymbol{H}\boldsymbol{C}\boldsymbol{\xi}(t) \end{aligned} \tag{12}$$

According to (12), the transfer function form of $\boldsymbol{r}$ is

$$\boldsymbol{r}(s) = \boldsymbol{T}_{rf}\boldsymbol{f}(s) = \boldsymbol{H}\boldsymbol{C}[s\boldsymbol{I} - (\boldsymbol{A} - \boldsymbol{R}\boldsymbol{C})]^{-1}\boldsymbol{B}_f \boldsymbol{f}(s) \tag{13}$$

Here, $\boldsymbol{T}_{rf}$ are the transfer functions from $\boldsymbol{f}$ to $\boldsymbol{r}$, and $\boldsymbol{R}$ and $\boldsymbol{H}$ are the matrices to be determined so that the residual r_i is only affected by the ith fault and decoupled from all other faults, i.e., $\boldsymbol{T}_{rf}$ is diagonal.

3. Multiple fault isolation framework

In this section, a fault isolation framework is proposed to isolate multiple micro thrusters faults of the drag-free satellite. Before doing so, a proportional-integral-derivative (PID) controller is presented to stabilize the drag-free system. Then, a bank of fault isolation filters is designed to achieve fault isolation of all the 12 FEEP thrusters. Note that the number of filters is much less than 12, which can save the on-board resources. The block diagram of the proposed fault isolation framework is shown in figure 3.

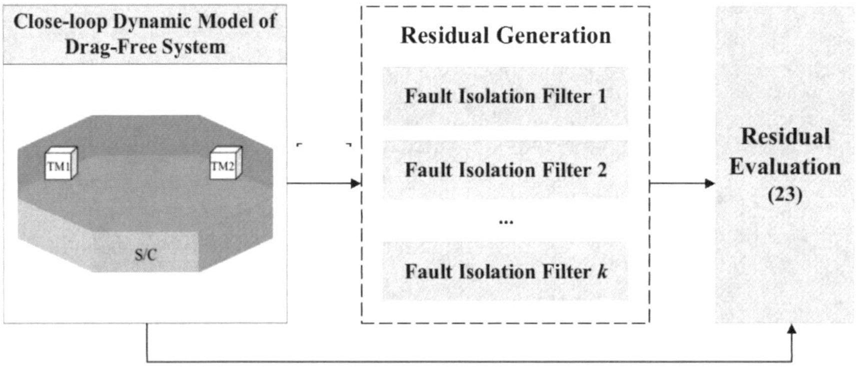

Figure 3. Block diagram of the proposed fault isolation framework.

3.1. Control laws

In this subsection, we design a fault isolation filter for the closed-loop system to prevent the subsequent introduction of control signals from affecting the fault isolation performance. To this end, three PID controllers are designed to stabilize the drag-free system. According to the science mode of the drag-free system, the control tasks can be divided into three parts by input decoupling:

1) Drag-free loop: the drag-free control DOF are fully controlled by the FEEP;

2) Suspension loop: the suspension DOF are fully controlled by the electrostatic actuator along the suspension axes;

3) Attitude loop: the satellite attitude DOF are controlled by the electrostatic actuator along the drag-free axes, then the drag-free loop will control the S/C to follow the drag-free coordinates.

According to the above description, the controllers can be expressed as

$$\begin{bmatrix} \boldsymbol{u}_{ES1} \\ \boldsymbol{u}_{ES2} \\ \boldsymbol{u}_{SC} \end{bmatrix} = \begin{bmatrix} \mathbf{0} & \mathbf{0} & \boldsymbol{K}_{DF} \\ \mathbf{0} & \boldsymbol{K}_{SUS} & \mathbf{0} \\ \boldsymbol{K}_{SC} & \mathbf{0} & \mathbf{0} \end{bmatrix} \begin{bmatrix} \bar{\boldsymbol{q}}_{DF} - \boldsymbol{q}_{DF} \\ \bar{\boldsymbol{q}}_{SUS} - \boldsymbol{q}_{SUS} \\ \bar{\boldsymbol{q}}_{SC} - \boldsymbol{q}_{SC} \end{bmatrix} \tag{14}$$

where $\boldsymbol{K}_{DF}$, $\boldsymbol{K}_{SUS}$ and $\boldsymbol{K}_{SC}$ represent the controller matrices, $\bar{\boldsymbol{q}}_{DF}$, $\bar{\boldsymbol{q}}_{SUS}$ and $\bar{\boldsymbol{q}}_{SC}$ are the reference inputs. The control input $\boldsymbol{u}_{SC} \in \mathbb{R}^{6\times1}$ is provided by 12 FEEP thrusters, i.e.

$$\boldsymbol{u}_{SC} = \boldsymbol{A}^{*}\boldsymbol{u}_{T} \tag{15}$$

where $\boldsymbol{A}^{*} \in \mathbb{R}^{6\times12}$ is the thrust allocation matrix, and $\boldsymbol{u}_{T} = [u_1, u_2, \cdots, u_{12}]^{\top} \in \mathbb{R}^{12\times1}$ is the desired thrust outputs of 12 thrusters.

3.2. Fault isolation framework

Consider the state space equation (12) and suppose the initial state error is zero, thus $\xi(t)$ can be expressed as

$$r_i(t) = \int_{t_0}^{t} e^{(A-RC)(t-\tau)} \boldsymbol{B}_{fi} f_i(\tau)\mathrm{d}\tau \tag{16}$$

where $\boldsymbol{B}_{fi}$ is the ith column of $\boldsymbol{B}_{f}$. For the aforementioned state-space dynamic model of the drag-free system, we can design fault isolation filter following the line of [9]. Before this, we should analyze the fault isolability of the drag-free system. According to [9], the fault detectability indexes $\boldsymbol{\rho}=[\rho_1,\rho_2,\cdots,\rho_{12}]$ is defined as

$$\rho_i := \min\{k : \boldsymbol{C}\boldsymbol{A}^{k-1}\boldsymbol{B}_{fi} \neq 0, k=1,2,\cdots\} \tag{17}$$

Based on this, the fault detectability matrix is defined as

$$\boldsymbol{D}=[\boldsymbol{C}\boldsymbol{A}^{\rho_1-1}\boldsymbol{B}_{f1},\ \boldsymbol{C}\boldsymbol{A}^{\rho_2-1}\boldsymbol{B}_{f2},\cdots,\ \boldsymbol{C}\boldsymbol{A}^{\rho_{12}-1}\boldsymbol{B}_{f12}] \tag{18}$$

Therefore, fault isolability can be determined by the rank of $\boldsymbol{D}$: if $\text{rank}(\boldsymbol{D})=m$, then the maximum number of fault that can be isolated is m.

According to the definition of the fault isolability, we are not sure to isolate 12 FEEP thrusters faults meanwhile by one fault isolation filter. Based on this situation, we design multiple sets of parallel running filters to isolate the entire fault set, and the fault subsets isolated by each fault isolation filter do not overlap with each other. For each fault isolation filter, the thrusters which are not isolated by this filter are considered to be fault free. Thus, the fault matrix $\boldsymbol{B}_f$ and fault signal $\boldsymbol{f}$ can be reorganized and divided into $\boldsymbol{r}$ subsets:

$$\begin{aligned} \boldsymbol{B}_f &\rightarrow \{\boldsymbol{E}_1,\boldsymbol{E}_2,\cdots,\boldsymbol{E}_r\} \\ \boldsymbol{f} &\rightarrow \{\boldsymbol{v}_1,\boldsymbol{v}_2,\cdots,\boldsymbol{v}_r\} \end{aligned} \tag{19}$$

where r is the total number of the fault isolation filters, and should be minimized while ensuring the isolation performance for the sake of conserving on-board resources; $\boldsymbol{E}_i$ and $\boldsymbol{v}_i$, $i=1,2,\cdots,r$ is composed of certain columns from $\boldsymbol{B}_f$ and $\boldsymbol{f}$. As such, the original system will also be partitioned into r subsystems

$$\begin{aligned} \dot{\boldsymbol{x}} &= \boldsymbol{A}\boldsymbol{x}+\boldsymbol{B}\boldsymbol{u}+\boldsymbol{E}_i\boldsymbol{v}_i \\ \boldsymbol{y} &= \boldsymbol{C}\boldsymbol{x} \end{aligned} \tag{20}$$

It is obvious that we can always find a set of subsets partitioning such that for each state space subsystem, the fault detectability matrix $\boldsymbol{D}$ is of full rank. Thus for each subsystem, the following theorem can be utilized to design fault isolation filter:

Theorem 1: Given system state space model as (20), and the form of fault isolation filter is

$$\begin{aligned} \dot{\hat{\boldsymbol{x}}} &= \boldsymbol{A}\hat{\boldsymbol{x}}+\boldsymbol{B}\boldsymbol{u}+\boldsymbol{R}_i(\boldsymbol{y}-\boldsymbol{C}\hat{\boldsymbol{x}}) \\ \hat{\boldsymbol{y}} &= \boldsymbol{C}\hat{\boldsymbol{x}} \\ \boldsymbol{r} &= \boldsymbol{H}_i(\boldsymbol{y}-\hat{\boldsymbol{y}}) \end{aligned} \tag{21}$$

thus the filter matrix $\boldsymbol{R}_i$ and $\boldsymbol{H}_i$ can be designed as

$$\begin{aligned} \boldsymbol{R}_i &= ([\boldsymbol{A}_{\rho_1}\boldsymbol{e}_1,\boldsymbol{A}_{\rho_2}\boldsymbol{e}_2,\cdots,\boldsymbol{A}_{\rho_k}\boldsymbol{e}_k-\boldsymbol{E}_i\boldsymbol{\Lambda}])\boldsymbol{D}^{+}+\boldsymbol{Z}_1(\boldsymbol{I}-\boldsymbol{D}\boldsymbol{D}^{+}) \\ \boldsymbol{H}_i &= \boldsymbol{W}\boldsymbol{D}^{+}+\boldsymbol{Z}_2(\boldsymbol{I}-\boldsymbol{D}\boldsymbol{D}^{+}) \end{aligned} \tag{22}$$

where $\boldsymbol{W}$ is any non-singular diagonal matrix, $\boldsymbol{Z}_1$ and $\boldsymbol{Z}_2$ are arbitrary matrices, $\boldsymbol{D}^{+}=(\boldsymbol{D}^{\top}\boldsymbol{D})^{-1}\boldsymbol{D}^{\top}$ is the Moore-Penrose inverse of $\boldsymbol{D}$, $\boldsymbol{\Lambda}$ is a diagonal matrix.

Proof: This proof is similar to [9] and is omitted here.

From the proof process of Theorem 1, it can be observed that if the filter parameters are chosen as (22), it is possible to make the transfer function matrix from $\boldsymbol{f}$ to $\boldsymbol{r}$ diagonal, thus the r sets of fault isolation filters can make each dimension of all filters residuals influenced by only one thruster fault signal, and the signals from the other thrusters are decoupled. Thus, the multiple faults isolation problem proposed in Section 2.3 is solved. Besides, a simple threshold logic is applied for the decision-making task, such as

$$\begin{cases} |\boldsymbol{r}(t)| < \delta, & \text{if fault-free} \\ |\boldsymbol{r}(t)| \geq \delta, & \text{if faulty} \end{cases} \tag{23}$$

where $\delta > 0$ is a user-defined threshold.

4. Numerical simulation

In this section, we conduct numerical simulations to showcase the efficacy of the fault isolation framework when applied to the drag-free system. We consider the presence of bias-type faults by default and perform simulations under the remaining three fault conditions as described in equation (7). The simulation step size is set to $0.001\mathrm{s}$, and the total simulation time is $50\mathrm{s}$. It's worth noting that only blocked-closed type faults are demonstrated here. According the practical drag-free dynamics, the fault detectability matrix satisfies

$$\operatorname{rank}(\boldsymbol{D}) = 6 \tag{24}$$

Thus, we need to design two fault isolation filters to isolation 12 FEEP thrusters.

Figure 4 shows the residuals behaviors of the 12 FEEP thrusters in both fault-free and faulty situations with the $4\mathrm{th}$, $8\mathrm{th}$ and $11\mathrm{th}$ thrusters closing on $25\mathrm{s}$. As expected, all the residuals keep below the threshold, while the $4\mathrm{th}$, $8\mathrm{th}$ and $11\mathrm{th}$ residuals exceed the threshold at $26.595\,\mathrm{s}$, $26.492\,\mathrm{s}$ and $26.942\,\mathrm{s}$, respectively, indicating that the corresponding three thrusters experienced faults. This demonstrates that the proposed fault isolation framework can effectively isolate multiple micro thruster faults in the drag-free system.

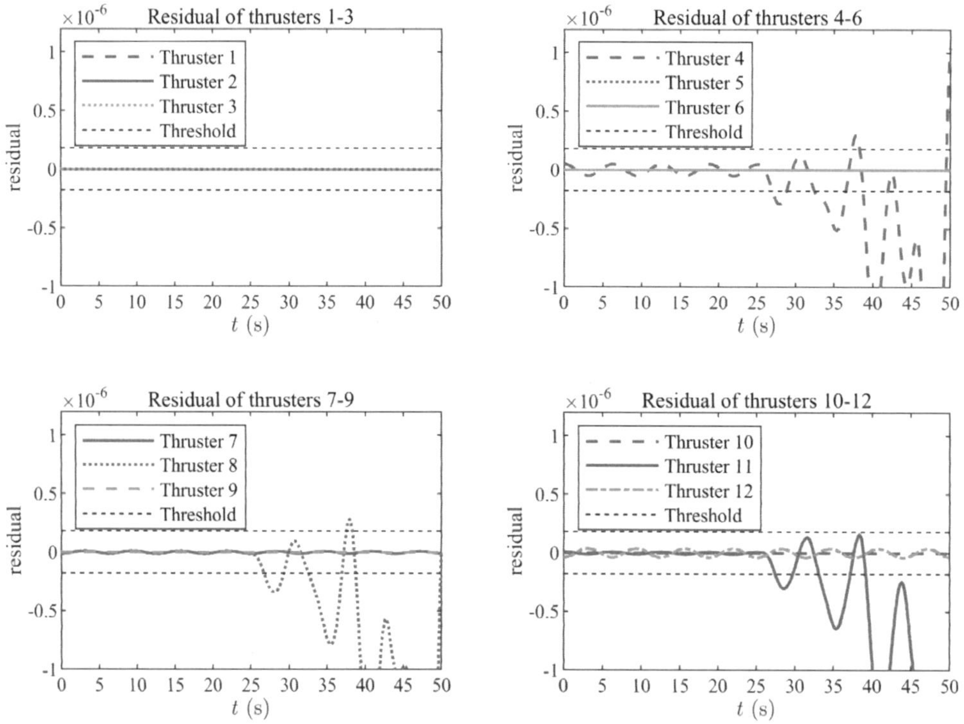

Figure 4. Residuals for the 12 FEEP thrusters.

5. Conclusion

This paper establishes a multiple faults isolation framework suitable for the micro thrusters faults of the drag-free system. Considering the unique structure and thrusters layout of the drag-free system, two sets of parallel fault isolation filters are designed to isolate all FEEP thrusters faults. The blocked-closed type faults of three thrusters are simulated to test its effectiveness, and the results shows that the framework can isolate faults occurring on the multiple thrusters, and all the faults can be isolated in $2\mathrm{s}$. Our future work will focus on the fault estimation problem based on the current framework.

Acknowledgments
This work was supported in part by the National Key R&D Program of China under Grants 2021YFC2202600 and 2021YFC2202603, and in part by Tianmushan Laboratory Research Project under Grants TK-2023-B-010 and TK-2023-C-020.

References

[1] C. Xiao, E. Canuto, W. Hong, Z. Zhou, and J. Luo, "Drag-free design based on embedded model control for tianqin project," *Acta Astronautica*, vol. 198, pp. 482-494, 2022.
[2] P. Gath, H. R. Schulte, D. Weise, and U. Johann, "Drag free and attitude control system design for the lisa science mode," *AIAA Guidance, Navigation and Control Conference and Exhibit*, p. 6731, 2007.
[3] A. Schleicher, T. Ziegler, R. Schubert, N. Brandt, P. Bergner, U. Johann, W. Fichter, and J. Grzymisch, "In-orbit performance of the lisa pathfinder dragfree and attitude control system," *CEAS Space Journal*, vol. 10, pp. 471-485, 2018.
[4] N. Tudoroiu and K. Khorasani, "Satellite fault diagnosis using a bank of interacting kalman filters," *IEEE Transactions on Aerospace and Electronic Systems*, vol. 43, no. 4, pp. 1334-1350, 2007.
[5] D. Henry, "Fault diagnosis of microscope satellite thrusters using H_∞ / H_- filters," *Journal of Guidance, Control, and Dynamics*, vol. 31, no. 3, pp. 699-711, 2008.
[6] S. X. Ding, "Fault isolation schemes," *Model-Based Fault Diagnosis Techniques: Design Schemes, Algorithms and Tools*, pp. 405-440, 2013.
[7] A. K. Prajapati and B. Roy, "Multi-fault diagnosis in three coupled tank system using unknown input observer," *IFAC-PapersOnLine*, vol. 49, no. 1, pp. 47-52, 2016.
[8] A. Falcoz, F. Boquet, G. Flandin, and P. Blanc-Paques, "Robust H_∞ / H_- thruster failure detection and isolation with application to the lisa pathfinder spacecraft," *AIAA guidance, navigation, and control conference*, p. 7906, 2010.
[9] B. Liu and J. Si, "Fault isolation filter design for linear time-invariant systems," *IEEE Transactions on Automatic Control*, vol. 42, no. 5, pp. 704-707, 1997.
[10] Z. Li and I. M. Jaimoukha, "Observer-based fault detection and isolation filter design for linear time-invariant systems," *International Journal of Control*, vol. 82, no. 1, pp. 171-182, 2009.
[11] W. Fichter, A. Schleicher, and S. Vitale, "Drag-free control design with cubic test masses," *Lasers, Clocks and Drag-Free Control: Exploration of Relativistic Gravity in Space*, pp. 361-378, 2008.
[12] R. Fonod, D. Henry, C. Charbonnel, and E. Bornschlegl, "Position and attitude model-based thruster fault diagnosis: A comparison study," *Journal of Guidance, Control, and Dynamics*, vol. 38, no. 6, pp. 1012-1026, 2015.

Journal of Physics: Conference Series **2633** (2023) 012014 doi:10.1088/1742-6596/2633/1/012014

Design and simulation of anti-sloshing baffles applied to detumbling payload propellant tanks

J C Tang[1], Z H Liang[1,*], K Zheng[1] and P Zheng[2]

[1] School of Mechanical Engineering, Nanjing University of Science and Technology, 200 Xiaolingwei Street, Nanjing, China.
[2] Shanghai Institute of Aerospace Systems Engineering, 3888 Yuanjiang Road, Shanghai, China.

* Corresponding author's email: liangzh2014@126.com

Abstract. The method to suppress the liquid sloshing in the tank applied to the jet detumbling payload is investigated in this study. Among all kinds of anti-sloshing structures, the symmetrical annular baffle was chosen due to the features of impact on the payload's tank. The damped mass-spring liquid sloshing model established is used to compare the damping ratio of the baffle with different baffle layers and different baffle heights. Using FLUENT software, based on VOF (volume of fluid) numerical simulation technology, the damping effect of baffles with different layers and heights is simulated. The results show that 3-layer 15mm or 4-layer 10mm baffles can effectively reduce the influence of liquid sloshing. The damping ratio of the storage tank with a 3-layer 15mm anti-sloshing baffle is at least 18.3 times that of a smooth tank, and the damping ratio of the tank with a 4-layer 10mm anti-sloshing baffle is at least 18.7 times that of a smooth tank.

1. Introduction

As the pace of human exploration of the universe gradually accelerates, space debris becomes a problem that cannot be ignored [1]. Based on ESA models, the true number of debris larger than 1 cm in size is likely over one million [2]. Most of them are tumbling due to their residual angular momentum [3]. As a result, detumbling process is required before using capture mechanisms like robotic arms and tentacles to capture space debris [4]. A variety of methods have been proposed to eliminate rotation, such as deceleration brush detumbling [5], mechanical pulse detumbling [6], tethered space robot detumbling [7], electrostatic detumbling [8], electromagnetic detumbling [9], gas shock detumbling[10], ion beam detumbling [11], etc. However, during the detumbling process, the free floating of the target makes it necessary for the service satellite to constantly adjust its attitude to carry on the detumbling operation. To solve this problem, jet detumbling payload is proposed. The detumbling payload is designed based on the principle of jet detumbling, launched from the service satellite, formed a rigid connection with the target through the attachment mechanism, and generated detumbling torque through the jet.

Considering the volume limitation of the service satellite, the jet propulsion system uses liquefied gas as the propellant. The actual storage state of the propellant is gas-liquid coexistence. The cold-gas propulsion system has been widely used in the orbital maneuvering task of micro-nano aircrafts. But due to the acceleration generated by the thruster is relatively small, the liquid sloshing in the propellant tank is rarely researched. The accelerations caused by the impact of the detumbling payload when launched from the service satellite and attached to the target surface are much bigger than the acceleration generated by the jet. The problem of propellant sloshing caused by external impact will directly affect the launch accuracy and attitude stability. Therefore, it is necessary to study the suppression of liquid sloshing in jet detumbling payload tanks. The commonly used anti-sloshing structures mainly include anti-sloshing baffle, flexible diaphragm and floater, among which the anti-sloshing baffle is more common. Maleki et al. developed an estimation of hydrodynamic damping ratio of liquid sloshing in baffled tanks undergoing horizontal excitation using Laplace's differential equation solution. The results implied that the ring baffles are more effective in reducing the sloshing oscillations [12]. Hasheminejad et al. studied the effects of liquid filling level, baffle arrangement, and length on liquid sloshing in horizontally placed cylindrical tanks [13], and the effect of baffles in the semi-filled tank under transverse excitation on liquid sloshing [14].Cho et al. designed a bottom-hinged, top-tensioned porous membrane baffle for rolling rectangular storage tanks and studied their performance as anti-sloshing devices [15]. Arun et al. studied the anti-sloshing effect of horizontal porous baffles in rectangular storage tanks [16]. Sobia et al. studied the use of slat screens to damp fluid sloshing during the refuel process of the tank under microgravity conditions. The images captured by the high-speed camera showed that the slat screen could effectively increase the sloshing damping [17]. Reza et al. compared the damping effect of the fixed baffle and the spring-connected movable baffle, and found that the double spring movable baffle can reduce the sloshing force on the sidewall by 30%~35% [18]. Both the floater method and the flexible diaphragm method aim to destroy the original free liquid surface to increase the energy dissipation. Bauer et al. proposed that if the free liquid surface of a viscous liquid is completely covered by an elastic structure, the damped natural frequencies will be larger than those obtained with a free liquid surface [19]. The coupled frequencies will increase with increasing liquid fill ratio and quickly reach a value that is close to constant [20]. Dillon et al. proposed a coupled fluid-structure CFD model to capture the damping effects of a flexible diaphragm on the propellant[21]. Koh et al. designed a constrained floating plate device [22]. It was demonstrated through numerical simulation and experiments that the floating plate can reduce the sloshing pressure of liquid on the tank. From 2010 to 2013, Anai Kim et al. designed a floating pad for large LNG storage tanks[23], which is made of melamine foam wrapped in marble, the quality can be determined by the size of the marble, and the flexural stiffness is controlled by the connection between the foams. Chongwei Zhang et al. studied the effect of the number of floating layers on liquid sloshing in storage tanks, and showed that as the number of layers increases, the amplitude of liquid level sloshing also decreases [24]. However, the formal anti-sloshing designs were aimed at large storage tanks, and set large external impacts and large shaking amplitudes. The structures designed were complex and difficult to process. As a result, they are not suitable for the detumbling payload proposed in this paper. The detumbling payload is a 10kg class micro-nano vehicle.

In this paper, impact conditions during the detumbling payload task are analyzed, and the damped mass-spring system model is used to simplify the linear simplification of liquid sloshing. Different forms of anti-sloshing method are compared, and a multi-layer symmetrical annular anti-sloshing baffle is designed. Taking the sloshing characteristic parameters in account, the VOF-based numerical simulation technology of Fluent software is used to describe the sloshing effect. Through simulation, the sloshing situation, sloshing damping and sloshing force of the propellant in different baffle designs are obtained, and the anti-sloshing baffle design suitable for jet detumbling payload is proposed by comparing various schemes.

2. Task background and sloshing suppression methods

2.1. Task background

The detumbling payload is carried on the service satellite, which is moored at a certain distance to the unstable target. After completing the observation of the target and stabilizing the attitude, the solid launch motor of the detumbling payload is ignited. Once hitting the target, the detumbling payload is fixed on it through the attachment mechanism, as shown in figure 1. Then, through the sensitive device carried by the detumbling payload, the angular rotational velocity of the unstable target is sensed. The propulsion system is controlled to reverse jet based on the designed detumbling control law to provide the torque to achieve rapid detumbling of the unstable target.

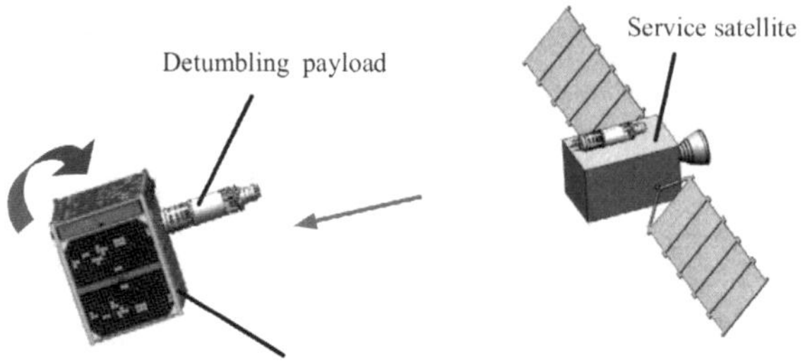

Figure 1. Detumbling payload mission process.

The detumbling payload is mainly composed of an attachment mechanism, a control module, Reaction Control System (RCS) thrusters, a launch motor, etc. The structure is shown in figure 2. As the core component of the detumbling payload, the cold-gas propulsion system provides a certain range of thrust and sufficient total impulse.

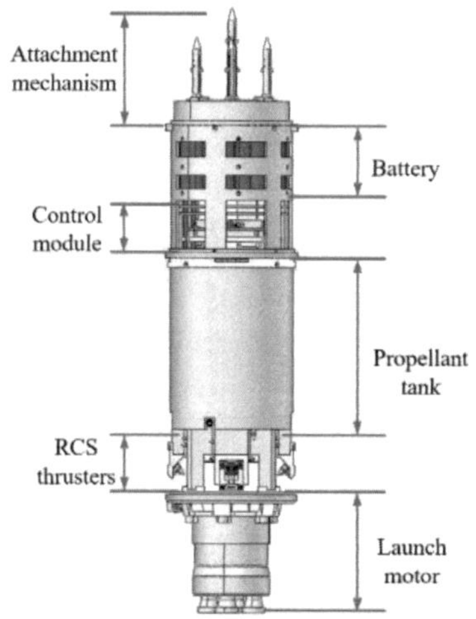

Figure 2. Structure of detumbling payload.

The payload is launched from a certain distance from the target, which means the pointing accuracy required is especially high. In the process of the payload approaching the unstable target the propellant will be impacted twice, as shown in figure 3. The first impact which directly acts on the storage tank is caused by the ignition of the launch motor. The second impact is caused by the attachment mechanism penetrating the unstable target, which causes the speed of the payload drops to zero instantaneously.

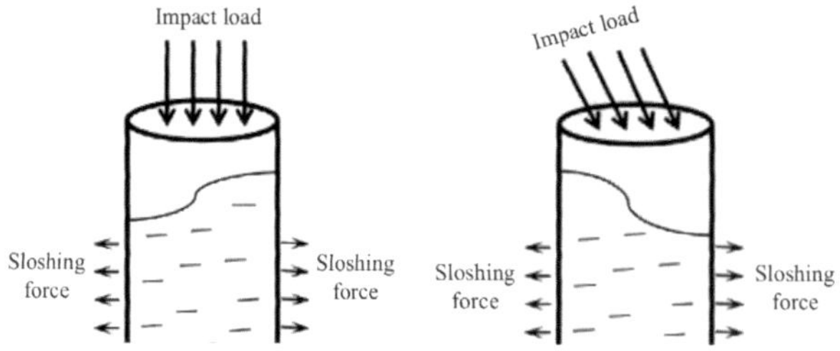

（a） launch impact （b） attachment impact

Figure 3. Two impact conditions.

2.2. Design of anti-sloshing method

2.2.1 Principle of anti-sloshing. Methods to describe liquid sloshing mainly include analytical theoretical method, numerical simulation calculation and experimental method. This paper will combine theory and simulation, start from the sloshing characteristic parameters, and use VOF-based numerical simulation technology to describe the sloshing effect.

The volume of fluid (VOF) method is widely used in the application field of liquid sloshing and free liquid flow. The VOF model is based on the N-S equation to represent the interface state between the two phases of surface, and the flow field is divided by the volume fraction ε_q of the q phase. The volume fraction of a single unit is calculated as follows:

$$\varepsilon_q = \frac{\iiint \varepsilon_q(x,y,z)dxdydz}{\iiint dxdydz} \quad (1)$$

In this design, there are only two phases, and under the premise of limited tank volume, the VOF model is used to calculate parameters such as sloshing damping.

Based on the theoretical analysis of the equivalent dynamic model, the analytical theory uses the damped mass-spring system or the equivalent pendulum model to study the dynamic effect of liquid sloshing, and makes a linear simplification [25]. The sloshing mode is described in equation (2):

$$K_r x_r + \dot{x}_r(K_r + C_r) = -m_r(\vec{a} + \vec{\alpha} \times \vec{r}) \cdot \vec{n} \quad (2)$$

Where x_r represents the displacement relative to the axis, r represents the sloshing mode of the first order, m_r corresponds to the sloshing mass, K_r is the equivalent stiffness of the rth order, C_r is the damping coefficient of the rth order. Based on the above formula, the calculation formula of the sloshing frequency ω_r, the sloshing damping ratio ζ_r, the shaking force F_r and the torque M_r can be further obtained:

$$\begin{cases} \omega_r = \sqrt{\frac{K_r}{m_r}} \\ \zeta_r = \frac{C_r}{2\sqrt{K_r m_r}} \\ F_r = m_r \dot{x}_r \\ M_r = m_r(h_r \dot{x}_r - g x_r) \end{cases} \quad (3)$$

where h_r is the height in the established tank coordinate system. It can be seen from experiments that the result project obtained after linear simplification is still available.

Therefore, the anti-sloshing design will start from the four sloshing characteristic parameters ω_r, ζ_r, m_r and h_r. The realization of the anti-sloshing function mainly depends on changing the free boundary conditions of the sloshing fluid surface or increasing the sloshing damping.

Changing the free fluid level boundary condition is to reduce the impact of sloshing by reducing the area of the fluid surface, changing the kinetic boundary condition, and absorbing the sloshing energy, as is shown in figure 4. However, this method will greatly reduce the amount of propellant. The commonly used diaphragm and float methods may be force-coupled with the storage tank, which greatly increases the local stress at the connection and causes problems such as ring deformation.

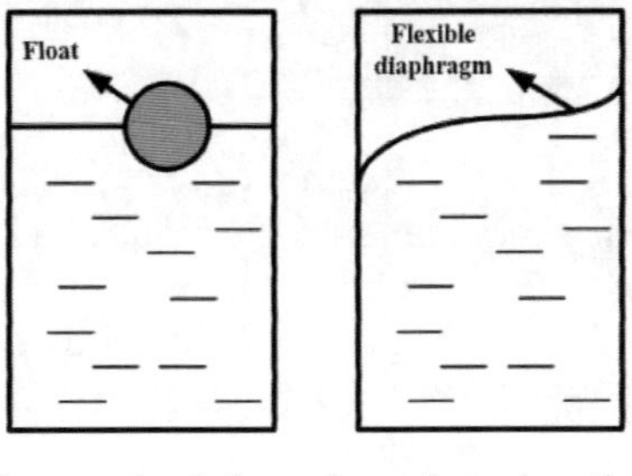

Figure 4. Anti-sloshing methods based on changing the boundary condition.

The main form of increasing sloshing damping is adding anti-sloshing baffles. Anti-sloshing baffles can also be processed with the tank by additive manufacturing, which is very suitable for the situation where the shape and size of the tank are difficult to adjust. For the detumbling payload, an anti-sloshing baffle will be set inside the tank to reduce the effect of liquid propellant sloshing.

For smooth storage tanks without anti-sloshing baffles, part of the damping comes from the viscosity of the liquid itself, and the viscosity of the liquid when sloshing produces a large energy dissipation in a thin layer of the inner wall of the tank, which is called the Stokes boundary layer. The other part comes from the energy dissipation caused by the viscosity of the remaining fluid. After adding anti-sloshing baffles, the boundary of the fluid is changed, which aggravates the energy dissipation of the above two types and produces winding damping. Because the relative speed of the storage tank and the liquid exists at the moment of sloshing, the baffle will hinder the flow of the liquid, resulting in different pressures on different sides of the baffle, and eventually creating passing flow resistance to reduce the sloshing of the liquid.

The dynamic viscosity coefficient of the propellant selected for this design is 3.0439×10^{-4}Pa · s (R236fa at 20°C). The dissipation of the Stokes boundary layer and the internal dissipation due to viscous flow are much smaller than the dissipation generated by passing flow resistance, so the passing flow damping is the only form of damping to be considered in the calculation.

The essence of the passing flow resistance is the pressure difference resistance. The pressure difference resistance hinders the liquid when sloshing and causes energy dissipation, which is given by:

$$\begin{cases} D_d = \rho \int_{S_w} C_d |\nabla\varphi \cdot \vec{n}|^3 \, dS \\ C_d = \frac{K}{\sqrt{\eta_{max}}} \end{cases} \tag{4}$$

where $\vec{n}$ is the unit normal vector of the designed anti-sloshing baffle, η_{max} is the maximum amplitude of shaking. K is related to the curvature of the baffle. For the plate, the value of K is 8.5. The Reynolds number will limit the value range of C_d.

The total mechanical energy of liquid sloshing in a cycle is:

$$E = \frac{\rho}{2} \int_{\Omega} |\nabla\varphi|^2 \, d\Omega \tag{5}$$

The damping ratio of this section can be calculated as:

$$\zeta_d = \frac{D_d}{2\omega E} \tag{6}$$

2.2.2 Design of anti-sloshing baffle. Commonly used anti-sloshing baffles are generally divided into three types: semicircular baffles, radial baffles and symmetrical ring baffles.

Semicircular baffle is typically a kind of asymmetric baffle, as shown in figure 5. Single-layer or multi-layer baffles are set in the tank There exists an angle between the central axis and the liquid sloshing direction. When the angle is bigger than 60°, the baffle doesn't work efficiently.

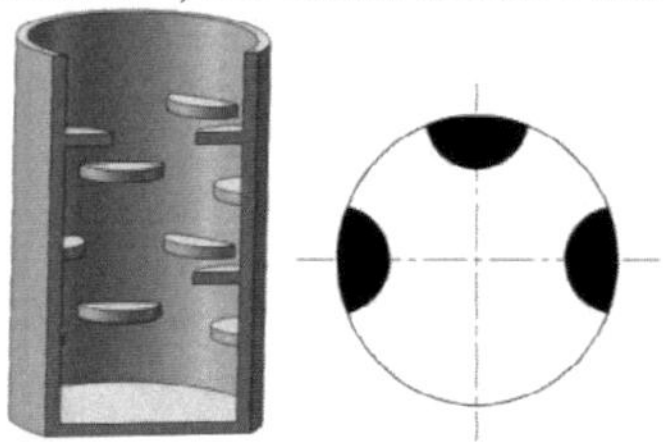

Figure 5. Schematic diagram of a semicircular baffle.

The higher damping efficiency of the semicircular baffle requires a larger radius r, and the efficiency is not completely positively correlated with the radius. Meanwhile the bearing capacity of the larger semi-circular baffle is poor, and the stiffness is much weaker than that of the annular baffle. Usually the empirical value range of radius is $0.2R\sim0.3R$. The damping ratio of this type of baffle can be calculated with the following empirical formula:

$$\zeta = KC\left(\frac{d}{R,\theta}\right)\alpha^k\left(\frac{\eta}{R}\right)^{0.5} \tag{7}$$

In the formula, k and K are both coefficients, the values are 2.83~3 and 1.5~1.8. The value of α depends on the specific design, which represents the ratio of the sum of the area of all semi-circular baffles to the cross-sectional area of the tank. $C\left(\frac{d}{R,\theta}\right)$ is generally obtained by fitting all the data. There is no fixed analytical expression.

Radial baffles, also known as vertical strip baffles, are evenly arranged along the circumference of the cross-sectional circumference of the tank, as shown in figure 6. The damping of this type can remain unchanged when the liquid level position changes, but will be affected by the change of sloshing direction. The design parameters of this type of baffle mainly include the width of the baffle w, the height of the baffle h, and the number of baffles n (the angle between adjacent baffles).

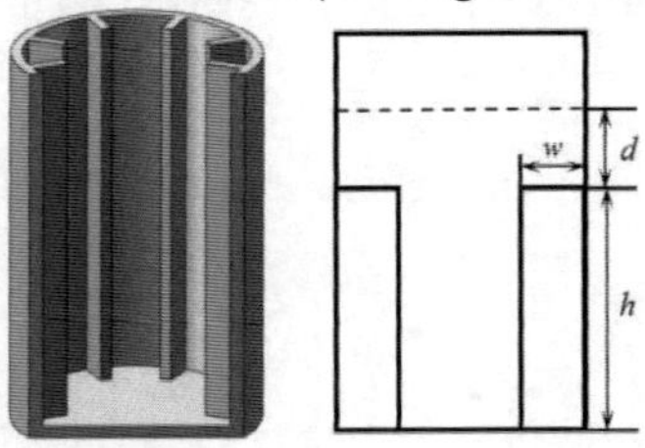

Figure 6. Schematic of the radial baffle.

For radial baffles, the height of the baffle represents the range it functions. With the number of baffles increases, the anti-sloshing effect will also increase. Meanwhile the mass will increase and the directionality of damping will be effected. In the case of normal filling ratio, the influence of damping caused by different shapes of the tank bottom is usually ignored. The damping ratio of this type of baffle can be calculated with the following empirical formula:

$$\begin{cases} \zeta = \frac{0.13}{1+\kappa}\left(\frac{w}{R}\right)^{1.5}\left(1-\frac{w}{2R}\right)^{-2.5}\left(\sum_{i=1}^{n}|cos\,\theta_i|^{2.5}\right)\left(\frac{\eta}{R}\right)^{0.5} \\ \kappa = 0.209\left(\frac{w}{R}\right)^2\left(1-\frac{w}{2R}\right)^{-2}\left(\sum_{i=1}^{n}cos^2\,\theta_i\right) \end{cases} \tag{8}$$

where η is the shaking width of the storage tank. θ_i is the angle between the first radial baffle and the direction of liquid sloshing, which is affected by the number of baffles.

The symmetrical annular baffle is generally a circular ring parallel to the liquid surface, as shown in 0. This type of baffle has the same damping in different directions. The thickness of the baffle is designed to be sufficient to consider the baffle as rigid baffle, the specific calculation formula is as follows:

$$F = \frac{\rho g R(1-V^2)}{E}\left(\frac{w}{R}\right)^5\left(\frac{R}{\tau}\right)^3 f\left(\frac{w}{R}\right) \tag{9}$$

where F is the flexibility parameter, ρ is the liquid density, R is the radius of the tank, V is the Poisson's ratio of the baffle processing material, corresponds to the modulus of elasticity E. w is the width of the baffle, and τ is the thickness of the baffle. $f\left(\frac{w}{R}\right)$ describes the effect of boundary conditions on the baffle. When the flexibility parameter is zero, the baffle is considered as rigid. The thickness of the baffle is designed according to the strength conditions, which can effectively improve the damping and anti-shake efficiency.

Considering the size of the tank, the anti-sloshing effect of the single-layer baffle is not ideal, so it is necessary to install a multi-layer baffle. The main design parameters are the width of the baffle w and the layer spacing s as shown in figure 7.

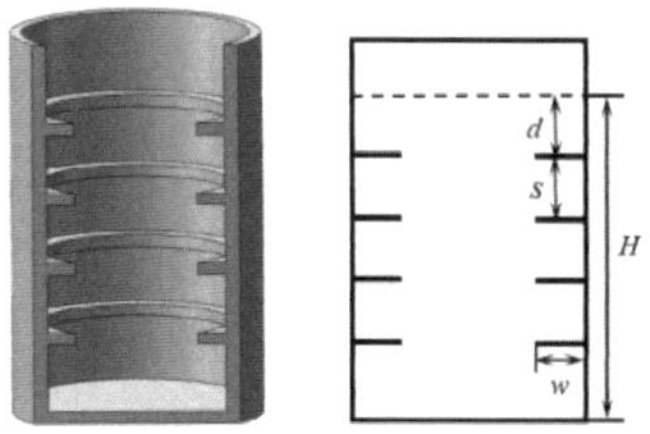

Figure 7. Schematic of the symmetrical annular baffles.

The distance between the annular baffle and the liquid level is d. When d is negative, the baffle exceeds the liquid level. The range of d which makes the baffle effective is $-0.1R \leq d \leq 0.45R$. Unless there is very little liquid in the storage tank, the anti-sloshing effect of the bottom baffle is not obvious. The distance between the baffles is designed as $0.2R \leq s \leq 0.4R$ to make the overall damping of the storage tank relatively uniform. In the primary stage of design, the damping of multi-layer baffles can be approximated as the linear superposition of anti-sloshing damping of multiple single-layer baffles. For the calculation of the damping ratio of symmetrical annular baffles, there is the following empirical formula:

$$\zeta = KC\left(\frac{d}{R}\right)\left[\frac{w\left(2-\frac{w}{R}\right)}{R}\right]^{1.5}\left(\frac{\eta}{R}\right)^{0.5} \tag{10}$$

where K is the coefficient, the value is 2.83~3. η is the width of the storage tank. There are different calculation expressions of $C\left(\frac{d}{R}\right)$ when the ratio of baffle width to the radius of the storage tank is different, as follows:

$$\begin{cases} 0.57 + \frac{5.7d}{R}, -0.1 \leq \frac{d}{R} \leq 0.07 \\ e^{\left(\frac{-4.6d}{R}\right)}, \frac{d}{R} \geq 0.07 \end{cases} \tag{11}$$

Since the impact conditions designed in this paper are all radial loads, the anti-sloshing effect of the radial baffle is not obvious, so the radial baffle is not selected. The propellant has complex sloshing behavior in the space environment without gravity, which requires the design of the baffle should be able to withstand large stress. Considering the processing method of additive manufacturing, the semicircular baffle is difficult to process and prone to stress concentration. So the semicircular baffle is not selected. Finally, a symmetrical ring baffle was chosen to suppress propellant sloshing in the tank.

Damping ratio is an important parameter to evaluate the anti-sloshing effect of the baffle. For the attenuation vibration system with anti-sloshing damping, the common vibration attenuation curve is shown in figure 8 below:

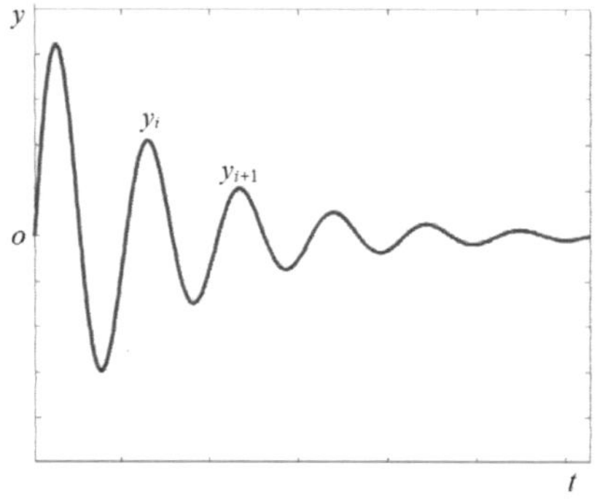

Figure 8. Attenuation vibration graph.

y_i and y_{i+1} in the figure are two adjacent vibration amplitudes, and the natural logarithm of the ratio of the two is called the logarithmic attenuation rate. The damping ratio is further solved by the following formula:

$$\zeta = \frac{1}{2\pi} ln \frac{y_i}{y_{i+1}} \quad (12)$$

The damping ratio obtained indicates the structural damping of the corresponding baffle. For structures with strict weight requirements, the ratio of damping value to baffle area can be further obtained, that is, the average damping ratio, which is used as a technical index to optimize the design.

2.3. Simulation model

In this paper, the algorithm SIMPLEC is selected for the sloshing characteristics of the impacted liquid, and the mesh motion conditions are added using UDF. The selected propellant R236fa has a density of $1377.5168 kg/m^3$, a dynamic viscosity coefficient of 3.0439×10^{-4}Pa · s, a surface tension coefficient of 0.0102N/m at 20°C and an internal pressure of 0.4MPa. The cylindrical tank used in simulation is shown in figure 9, with a size of Φ150mm×200mm and a wall thickness of 3mm. The coordinate origin is centered on the base, and the Z axis coincides with the cylindrical axis.

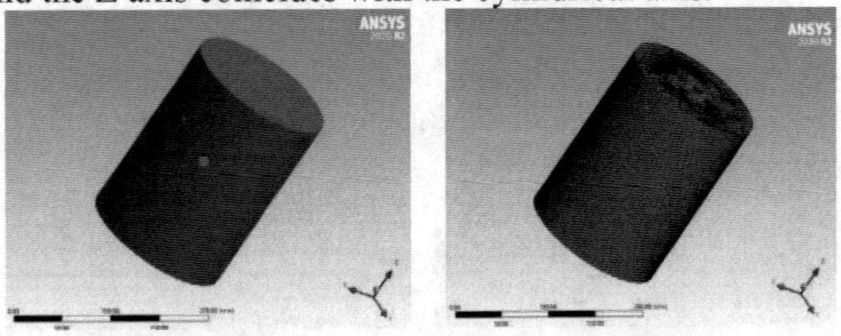

Figure 9. Storage tank model and meshing.

Different filling ratios are set with the internal pressure unchanged. The filling ratio is set to 50%, 75% and 90% to describe the sloshing situation under the two working conditions.

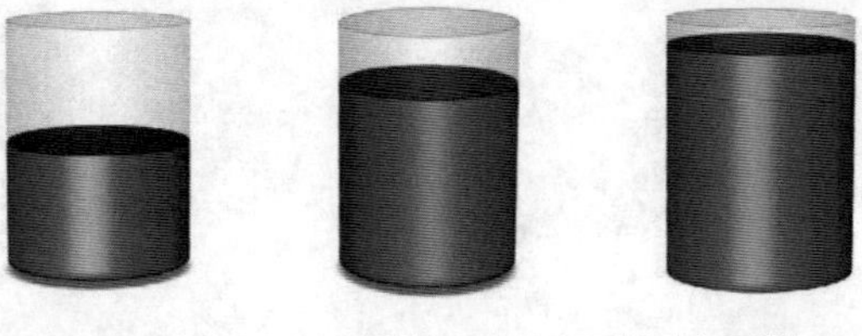

(a) Filling ratio 50% (b) Filling ratio 75% (c) Filling ratio 90%

Figure 10. Simulation models with different filling ratios.

The main design parameters of the ring baffle are the width of the baffle w and the layer spacing s. The thickness of the baffle only needs to consider the stress situation inside the tank. Using equation (9), the thickness is calculated to be 5mm. the layer spacing s characterizes the range of action of the single-layer baffle. The range of action given by the empirical formula is $0.25R \leq s \leq 0.55R$, that is, between 18.75mm ~ 41.25mm. The sloshing generated by the propellant at the bottom of the tank has little effect on the entire load. In order to achieve the optimal layer spacing, set the width of the baffle to 10mm unchanged, respectively draw the 3-layer baffle the 4-layer baffle and 5-layer baffle. Models are shown in the figures below after importing Fluent:

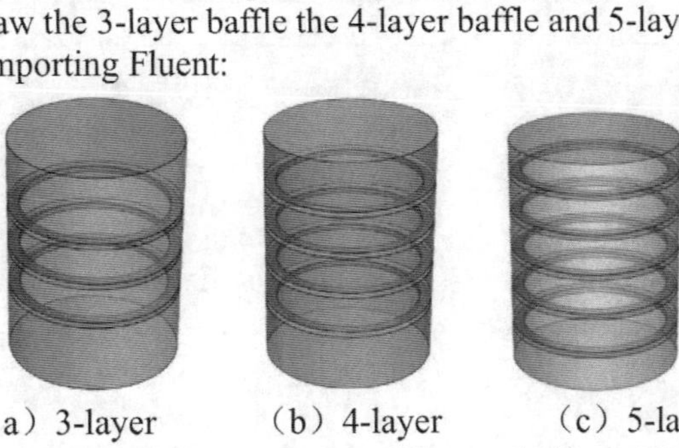

(a) 3-layer (b) 4-layer (c) 5-layer

Figure 11. The different number of layers baffle in FLUENT.

The tank and the liquid propellant can be considered as a mechanical system with a free liquid level. During the mission, the launch impact produces a strong nonlinear effect. Although the sloshing time is short, it has a huge influence on accuracy. The sloshing of the propellant caused by the attachment impact may cause lateral vibration of the payload, resulting in unstable attitude control. The specific impact values set are shown in the following table:

Table 1. Specific impact condition

	Starting speed(m/s)	End speed(m/s)	Duration (ms)	Angle between the impact and the Z axis
Launch	0	30	100	0°
Attachment	30	0	100	30°

3. Numerical simulation analysis

Taking the impact condition of the launch as the initial condition, the tank with three different numbers of layers were simulated and analyzed. The liquid-filling ratio of 50% was set. The first contact with the top after the impact was used as a control and the following results were obtained by simulation:

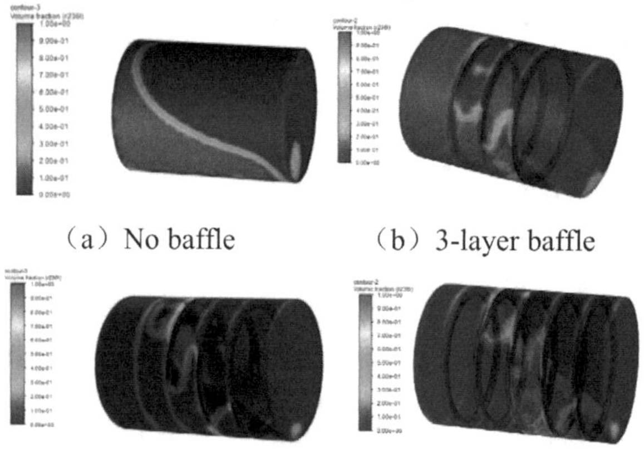

（a） No baffle （b） 3-layer baffle

（c） 4-layer baffle （d） 5-layer baffle

Figure 12. Sloshing conditions first time over the top.

As shown in figure 12, tank with a 3-layer baffle has an obvious damping effect compared to the smooth tank. The damping effect of the 4-layer baffle is significantly improved compared to the 3-layer baffle. Compared with the 4-layer baffle, the 5-layer baffle does not improve the anti-sloshing effect obviously. The first topping time of the three types of tanks with baffles and tanks without baffles is compared, the specific time is shown in the following table:

Table 2. First Topping time of tanks with different baffles

	No baffle	3-layer	4-layer	5-layer
Topping time(s)	0.47	0.86	1.35	1.51

It can be seen that the topping time of tanks with no baffle, 3-layer baffle, 4-layer baffle, and 5-layer baffle increases sequentially, indicating that the baffle has an anti-sloshing effect. The topping time extension of the 5-layer tank is relatively small compared to the 4-layer tank, indicating that the increase of anti-sloshing effect is not obvious. Taking the 4-layer baffle with a 50% filling ratio as an example, the sloshing force on tank's inner wall was analyzed, as shown in figures below:

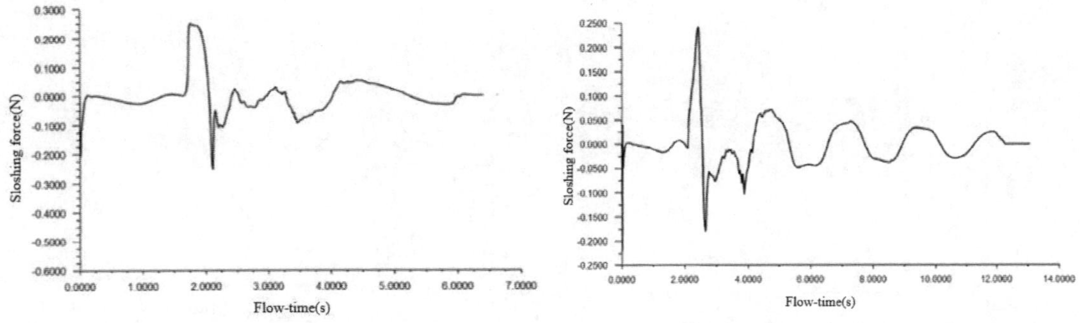

(a) 4-layer baffle (b) no baffle

Figure 13. Sloshing force of tank with 4-layer baffle and no baffle under a filling ratio of 50%.

Compared with the smooth tank under the same circumstances, it can be seen that the overall sloshing time becomes shorter, the sloshing force decreases. The damping of tank can be obtained under numerical simulation using equation (12). Meanwhile the damping calculation expression of the ring baffle is given in equation (10), and the damping ratio of different baffles in attachment impact condition can be obtained, and the specific result comparison is as follows:

Table 3. Damping ratios for different number of layers.

	Filling ratio 50%	Filling ratio 75%	Filling ratio 90%
No baffle Analytical solution	0.04329	0.04179	0.04142
No baffle Launch impact	0.04478	0.03997	0.03894
No baffle Attachment impact	0.04367	0.04035	0.03913
1-layer baffle Analytical solution	0.2082	0.1475	0.0933
3-layer baffle Launch impact	0.6172	0.4437	0.2774
3-layer baffle Attachment impact	0.6154	0.4421	0.2705
4-layer baffle Launch impact	0.8348	0.5898	0.3815
4-layer baffle Attachment impact	0.8325	0.5887	0.3784
5-layer baffle Launch impact	0.8735	0.6195	0.3957
5-layer baffle Attachment impact	0.8699	0.6128	0.3902

It can be seen from the simulation results that the damping ratio under different filling ratios will change. The analytical solution of the damping ratio of the tank without baffles is close to the simulated value, which proves that the simulation is credible. Compared with smooth tank, 3-layer baffle can improve the damping ratio by about 4.65 times. The damping ratio in launch condition is relatively larger than that in attachment condition. The dot line diagrams are shown below:

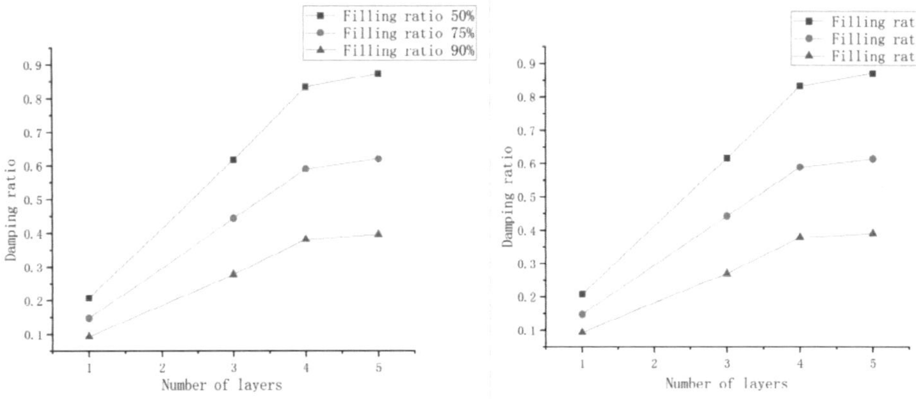

(a) launch condition (b) attachment condition

Figure 14. Damping ratio of baffles in launch condition and attachment condition.

The damping ratio of the baffle should theoretically be linearly superimposed with the increase of the number of layers. It can be seen from the simulation results that the damping ratio of the three and four layers is almost three and four times that of the single layer, but the 5-layer baffle does not increase the damping ratio according to this law. The increase of the baffle will increase the mass of the overall structure, so the priority is given to the design of the baffle for 3-layers (layer spacing 40mm) or 4-layers (layer spacing 30mm).

Literature [26] concludes that the anti-sloshing effect increases with the increase of the width of the baffle, and gives the following experience formula:

$$\overline{S} = \frac{S_d}{S_y = \frac{w(2R-w)}{R^2}} \leq 0.47 \tag{13}$$

S_d and S_y are the area of the baffle and the area of the free liquid level. The radius of the chamber of the storage tank is known as $R = 75\text{mm}$, so it can be solved that $w \leq 20.39\text{mm}$. Taking 50% filling ratio as an example, tests were performed to obtain the optimal baffle width.

Table 4. Baffle design schemes.

Number of layers	Width of baffle(mm)
3	10
3	15
3	20
4	10
4	15
4	20

Now taking the attachment condition as the initial condition, in which the 3-layer 10mm baffle and 4-layer 10mm baffle scheme have been analyzed in the above baffle layer spacing design process and obtained the damping ratio. The remaining four schemes are analyzed now, and the sloshing situation at the same moment is shown in figures below:

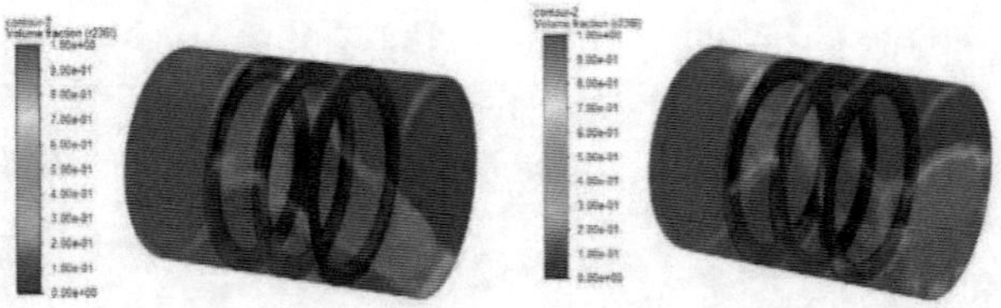

(a) 3-layer 15mm baffle

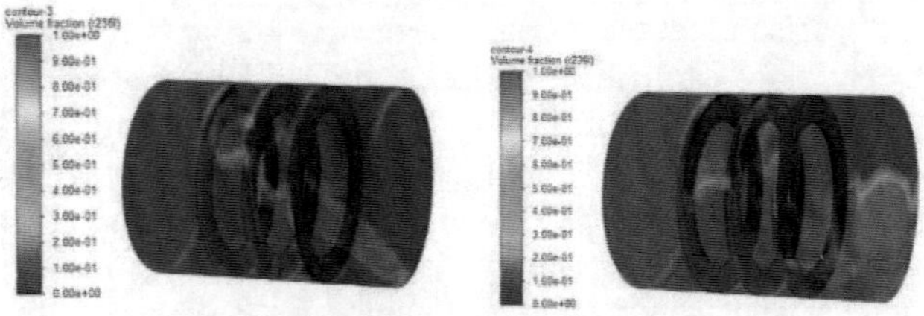

(b) 3-layer 20mm baffle

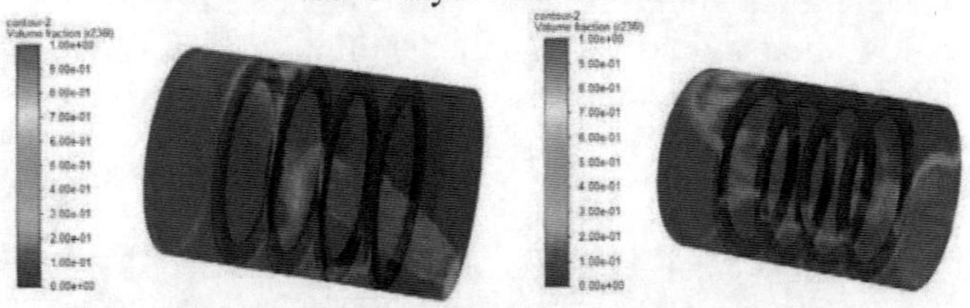

(c) 4-layer 15mm baffle

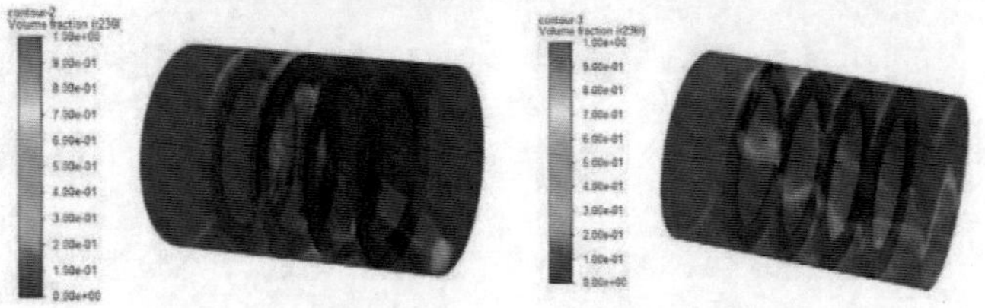

(d) 4-layer 20mm baffle

Figure 15. Sloshing situation in different designs.

From the sloshing situation simulation, it can be seen that the anti-sloshing effect of the 3-layer and 4-layer baffles increases with the increasing width of the baffle. But the effect increase is not as large as the width increase. Taking the force on the side wall of the tank with 3-layer 15mm baffle as an example, the simulation curve is shown in the figure below:

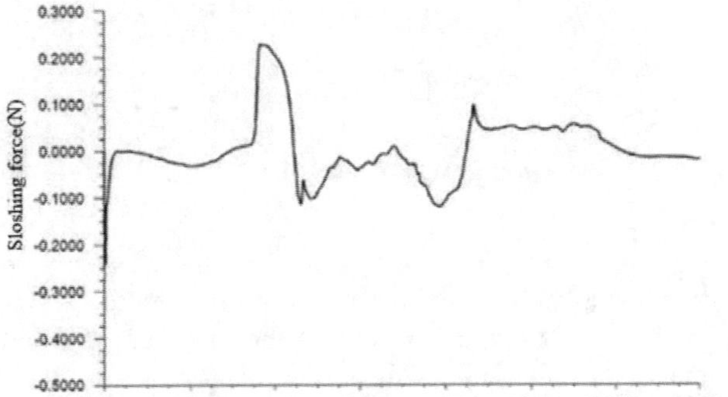

Figure 16. Sidewall shaking power of a 3-layer 15mm baffle storage tank.

From this curve, the damping ratio can be calculated. Similarly, the damping ratios of the other schemes in 0can be calculated. Because the density is certain, the volume of the baffle can reflect the weight. The results of each scheme are shown in the following table:

Table 5. Damping ratio and volume of each scheme.

	3-layer 10mm baffle	3-layer 15mm baffle	3-layer 20mm baffle	4-layer 10mm baffle	4-layer 15mm baffle	4-layer 20mm baffle
Launch	0.6172	0.8213	0.8466	0.8368	0.8451	0.8575
Attachment	0.6154	0.8429	0.8522	0.8325	0.8504	0.8746
Volume (cm^3)	68	94	120	91	127	160

It can be seen from the comparison that the increase in the number of baffle layers is more suitable for the impact of the launch condition, while the increase of the width of the baffle is more suitable for the attachment condition. Although the mass of the 3-layer 10mm baffle is relatively small, the damping ratio is small as well. The 20mm width scheme has the highest damping ratio, while the mass is correspondingly larger. With the huge increase in baffle width, the 4-layer 15mm baffle is not greatly improved in damping ratio. In summary, the 3-layer 15mm and 4-layer 10mm schemes are more suitable for this design, of which the 3-layer 15mm is more suitable for the impact of the attachment condition, and the 4-layer 10mm is more suitable for the impact of the launch condition.

Finally, it is known that 75% filling ratio is in line with the filling situation in the actual project. In order to further verify the effectiveness of the symmetrical ring baffles, in the case of a filling ratio of 75%, the sloshing of the tank without a baffle (left) and the sloshing of the 4-layer 10mm baffle (right) are compared, as shown in figures below:

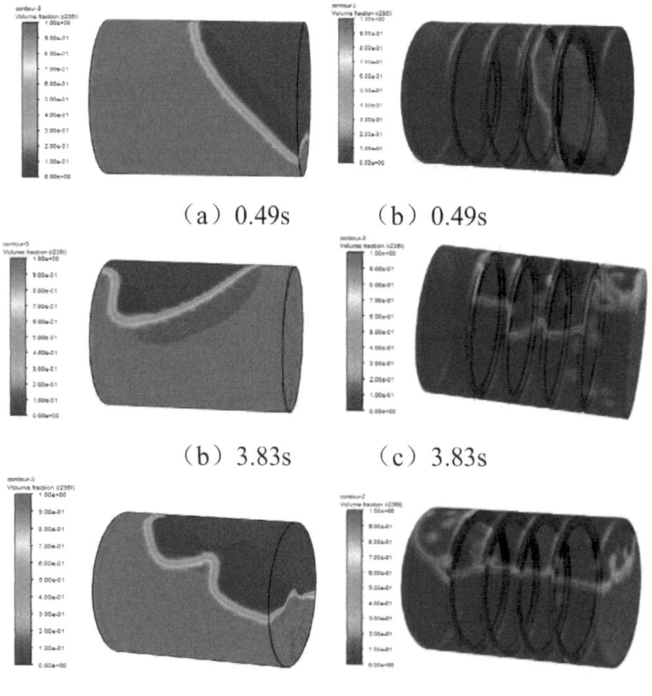

(a) 0.49s (b) 0.49s

(b) 3.83s (c) 3.83s

(d) 6.23s (e) 6.23s

Figure 17. Sloshing conditions with a filling ratio of 75%.

It can be seen from the simulation that the sloshing of the tank with the baffle at each moment is significantly weaker than that of the tank without the baffle, which confirms that the symmetrical annular baffle can effectively prevent sloshing, and quickly and effectively suppress the sloshing after the two impacts, ensuring the pointing accuracy of the detumbling payload and the stability of attitude control.

4. Conclusion

In this paper, the jet detumbling payload propellant tank is taken as the research object. The two impacts when launching and hitting the unstable target are selected as the research conditions. The liquid sloshing characteristic parameters are analyzed by the equivalent kinetic model. The anti-sloshing design is proposed by increasing the sloshing damping. Comparing different anti-sloshing baffle forms, considering the limitations of processing technology and actual working conditions, the symmetrical annular baffle was finally selected as the anti-sloshing baffle form. The parameter design of the baffle is carried out, the control variables of the width and number of layers of the baffle are numerically simulated, and the Fluent simulation conditions are set based on the liquid sloshing mechanism. Finally two baffle design schemes are selected, namely the 3-layer (layer spacing 40mm) 15mm baffle and the 4-layer (layer spacing 30mm) 10mm baffle. The damping ratio of the storage tank with a 3-layer 15mm anti-sloshing baffle is at least 18.3 times that of a smooth tank, and the damping ratio of a storage box with a 4-layer 10mm anti-sloshing baffle is at least 18.7 times that of a smooth tank. The simulation results show that the overall sloshing time of the propellant becomes shorter and the sloshing force decreases after adding the anti-sloshing baffle, which verifies the effectiveness of anti-sloshing baffle.

References

[1] Tang E L, Xie C R, Guo K, Han Y F, Chen C , Chang M A and He L P 2023 Analysis of space debris impact source localization based on PVDF piezoelectric film *Adv. Space Res.* **71(1)** 390–407

[2] ESA's Annual Space Environment Report Produced with the DISCOS Database 2022 *ESA Space Debris Office.* 8–18

[3] Shan M H, Guo J and Gill E 2016 Review and comparison of active space debris capturing and removal methods *Prog. Aerosp. Sci.* **80** 18–32

[4] Liu Y Q, Liu X F, Cai G P, Xu F and Tang S Y 2022 Detumbling a Non-Cooperative Tumbling Target Using a Low-Thrust Device *AIAA J.* **60** 1–12

[5] Nishida S I, Kawamoto S 2011 Strategy for caturing of a tumbling space debris *Acta Astronaut.* **68** 113–130

[6] Matunaga S, Kanzawa T, Ohkami Y 2001 Rotational motion-damper for the capture of an uncontrolled floating satellite *Control. Eng. Pract.* **9** 199–205

[7] Huang P F, Zhang F, Cai J, Wang D K, Meng Z J and Guo J 2107 Dexterous tethered space robot: design, measurement, control, and experiment *IEEE T Aero. Elec. Sys.* **53** 1452–68

[8] Bennett T and Schaub H 2015 Touchless electrostatic three-dimensional detumbling of large axi-symmetric debris *J Astronaut Sci.* **62** 233–253

[9] Gómez N O and Walker S J I 2015 Earth's gravity gradient and eddy currents effects on the rotational dynamics of space debris objects: Envisat case study *Adv. Space Res.* **56** 494–508

[10] Peters T V and Olmos D E 2016 COBRA contactless detumbling *CEAS Space Journal.* **8** 143–165

[11] Bombardelli C and Pelaez J 2011 Ion beam shepherd for contactless space debris removal *J. Guid. Control Dyn.* **34** 916–920

[12] Maleki A and Ziyaeifar M 2008 Sloshing damping in cylindrical liquid storage tanks with baffles *J. Sound Vib.* **311** 372–385

[13] Hasheminejad S M and Mohammadi M M 2011 Effect of anti-slosh baffles on free liquid oscillations in partially filled horizontal circular tanks *Ocean Eng.* **38** 49–62

[14] Hasheminejad S M, Mohammadi M M and Jarrahi M 2014 Liquid sloshing in partly-filled laterally-excited circular tanks equipped with baffles *J. Fluids Struc.* **44** 97–114

[15] Cho I H and Kim M H 2020 Effect of a bottom-hinged, top-tensioned porous membrane baffle on the sloshing reduction in a rectangular tank *Appl. Ocean Res.* **104** 100345

[16] George A and Cho I H 2021 Anti-slosh effect of a horizontal porous baffle in a swaying/rolling rectangular tank: Analytical and experimental approaches *Int. J. Nav. Arch. Ocean.* **13** 833–847

[17] Nadeem S, Bosca P, Stramenga M, Casella J, Coppens J, Henriques J, Wynn G and Hamed M 2022 An experimental investigation of slat-screens to mitigate fluid sloshing in microgravity

J. Space Sa.f Eng. **9** 319–327

[18] Saghi R, Ning D, Mayon R and Zhao M 2011 Evaluation of angular and linear spring systems linked to a dual baffle against the sloshing phenomenon *Ocean Eng*. **266** 113045

[19] Bauer H F and Chiba M 2005 Axisymmetric oscillation of a viscous liquid covered by an elastic structure *J. Sound Vib.* **281** 835–847,

[20] Bauer H F and Chiba M 2007 Viscous oscillations in a circular cylindrical tank with elastic surface cover *J. Sound Vib.* **304** 1–17,

[21] Sances D, Gangadharan S, Sudermann J and Marsell B 2010 CFD Fuel Slosh Modeling of Fluid–Structure Interaction in Spacecraft Propellant Tanks with Diaphragms *NASA Technical Reports*. 5–19

[22] Koh C G, Luo M, Gao M and Bai W 2013 Modelling of Liquid Sloshing with Constrained Floating Baffle *Comput. Struct.* **122** 270–279.

[23] Anai Y, Ando T, Watanabe N, Murakami C and Tanaka Y 2010 Development of a New Reduction Device of Sloshing Load in Tank *The Twentieth Int. Offshore and Polar Eng. Conf. (Beijing, China)* 18–26

[24] Zhang C W, Su P and Ning D Z 2019 Hydrodynamic study of an anti-sloshing technique using floating foams *Ocean Eng.* **175** 62–70.

[25] Farid M, Levy N and Gendelman O V 2017 Vibration mitigation in partially liquid-filled vessel using passive energy absorbers *J. Sound Vib.* **406** 51–73.

[26] Dodge F T 2020 The new" dynamic behavior of liquids in moving containers" *TX: Southwest Research Inst.(San Antonio)*

[27] Shimanovsky A O, Kuzniatsova M G, Yakubovich V I 2018 Dynamics of Tank Trucks with Baffles for Transportation of Viscous Liquids *Int. J. Rob. Res.* **7** 438-443

Simulation and analysis of delamination damage propagation in CFRP structures

Peiyuan Li [1], Qiaoyan Cai [2,*], Fei Wang [1], Tao Zhang [1] and Riming Tan[3]

[1] Beijing Institute of Astronautical Systems Engineering, Nandahongmen Rd. 1, 100076 Beijing, China

[2] China Academy of Launch Vehicle Technology, Nandahongmen Rd. 1, 100076 Beijing, China

[3] Beijing Institute of Space Launch Technology, Nandahongmen Rd. 1, 100076 Beijing, China

*** Corresponding author's email: aseexplorer@163.com**

Abstract. For aerospace vehicles, composite materials are used in large-scale integral panels due to their excellent material properties. In the structure of vehicles, this kind of composite material panel is connected to the internal structure through fasteners. During the composite material molding process and the process of drilling connection holes, some local delamination sometimes occurs, this delamination is hardly detected in time, it can cause great damage to the compressive strength of the structure. In this paper, the ABAQUS/Explicit solver is introduced to analyze the progressive damage of interlaminar separation with cohesive elements, the accuracy difference between different damage modes with typical solvers is verified. Then based on the self-compiled three-dimensional Hashin failure criterion subroutine, the numerical simulation is carried out on the compressive strength of the composite laminate under different pre-delamination damage ranges, and the compressive strength of the single-fastener connection structure with pre-delamination, indicates the fiber damage, matrix damage and delamination damage, in order to explore its failure mode and residual strength law.

1. Introduction

For aerospace vehicles, composite materials are used in large-scale integral panels due to their excellent material properties. In the structure of vehicles, this kind of composite material panel is connected to the internal structure through fasteners. During the composite material molding process and the process of drilling connection holes, some local delamination sometimes occurs, this delamination is hardly detected in time, it can cause great damage to the compressive strength of the structure. In order to evaluate the effect of this layered destruction, some scholars have done a lot of work through numerical methods and experiments. Fu et al.[1] studied the delamination damage mechanism and prediction method of composite laminates, established a macro-micro finite element model of orthogonal laminates that focused on the resin enrichment area between layers. Based on the three-dimensional Hashin failure criterion, the quadratic nominal stress criterion and the BK failure criterion in the bilinear Cohesive damage model, Zhang[2]used the VUMAT subroutine in the ABAQUS/Explicit module to simulate delamination damage during drilling of carbon fiber composite unidirectional laminates. Based on the research on the delamination failure mechanism of composite materials based on the cohesion theory, Gu[3] proposed a cohesive force/friction contact coupling model to explore the influence of friction coefficient, cohesive force strength, ply angle, etc. on friction delamination and load response. Xiao[4] and Jia[5] explored the residual strength of damaged composite laminates. Wu[6] and Zhang[7] conducted numerical simulation research on the propagation behavior of delamination defects. Zhao et al.[8] simulated the mode I and mode I/II delamination expansion of multidirectional laminates based on the cohesion model. Maruschak et al.[9] analyzed the initiation process of metal cracks from the quasi-brittle stage to fatigue cracks from a microscopic level.

Composite material structures will face internal interlayer damage caused by low-speed impacts during use. During the drilling process of connection holes, the stress-sensitive areas around the holes will also face the same problem. Evaluation of delamination effects for different initial delamination ranges and analysis of damage progression help to evaluate the remaining mechanical properties of the structure. Although the above results partially introduce explicit solvers, they do not discuss the scope of application of the explicit solver in different modes of damage, and the error analysis with the implicit solver. For the orthogonal unidirectional tape-laminated carbon fiber reinforced composite materials and composite material connecting joints used in a specific project, the damage effects under different damage effects also need to be discussed. In this paper, the ABAQUS/Explict solver is introduced to analyze the progressive damage of interlaminar separation with cohesive elements, the accuracy difference between different damage modes with typical solvers is verified in section 2. Then combined with the self-compiled three-dimensional Hashin failure criterion subroutine, and introduce defining an independent reduction factor for each damage mode, the reduction factor is directly multiplied by the material stiffness coefficient to establish a progressive damage model suitable for low-velocity impact of carbon fiber reinforced composite laminates. the numerical simulation is carried out on the compressive strength of the composite plate under different pre-delamination damage ranges, and the compressive strength of the single-fastener connection structure with pre-delamination, indicates the fiber damage, matrix damage and delamination damage, in order to explore its failure mode and residual strength law. In the last section, based on the Hashin failure criterion, the compressive strength of single-nail connected composite laminated panels with preset delamination is simulated and analyzed, and the failure modes around the connection holes are observed.

2. Comparison of explicit and implicit calculation results of three model of cracks

The ABAQUS/Explict solver was originally used to simulate transient problems such as impact and collisions. Compared with the ABAQUS/Standard solver, it is more conducive to solving complex contact problems, avoiding non-convergence problems in implicit solutions, require less computing power[10]. When using the ABAQUS/Explict solver to analyze quasi-static problems, a longer analysis step time and a smooth amplitude curve are used to slow the loading process, so that the dynamic effects in the model are not obvious, and the results close to the static analysis are obtained as much as possible. Generally, it can be considered that the calculation accuracy of the ratio of kinetic energy to internal energy of the system is less than 5%, which is acceptable for engineering.

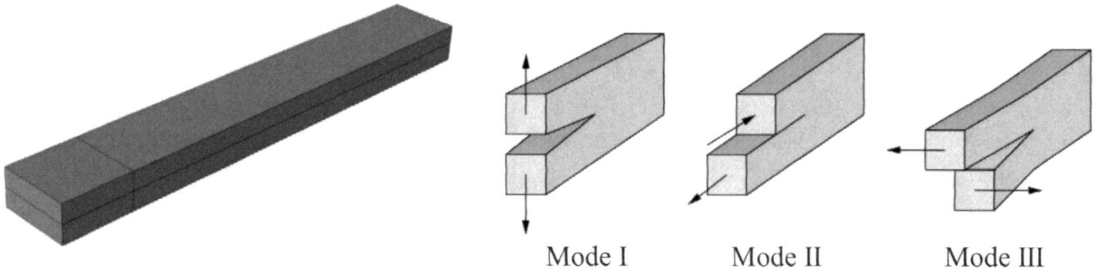

Figure 1. Simulation model and different modes of damage.

In order to verify the similarities and differences between the explicit solution and the implicit solution of ABAQUS in typical mode of cracks growths, the finite element calculation and analysis of three different fracture modes are specially selected for delamination evolution examples. As shown in the figure, two thick metal plates which have sufficient rigidity stacked up and down. The head end of the two plates is a free end, which is a square preset layer inside. The boundary condition of the plate is four-sided hinged support on the bottom surface (U_1=U_2=U_3=0), the free end surface is coupled a reference point, and the displacement in the corresponding direction is applied according to different damage types.

Cohesive elements damage is divided into two stages: damage initiation and damage evolution. In this example, the BK failure criterion is used to predict the damage evolution, and the quadratic nominal stress criterion is used to define the damage initiation stage. the expression is as follows[11]:

$$\left\{\frac{\langle\sigma_n\rangle}{\sigma_n^0}\right\}^2+\left\{\frac{\sigma_s}{\sigma_s^0}\right\}^2+\left\{\frac{\sigma_t}{\sigma_t^0}\right\}^2=1 \tag{1}$$

In the formula, σ_n is the normal stress in the thickness direction, $\langle\sigma_n\rangle$ is the Macaulay Bracket formula of σ_n, σ_s, and σ_t are the shear stress parallel to and perpendicular to the thickness direction, σ_n^0, σ_s^0 and σ_t^0 are the maximum nominal stresses of the cohesion model under pure tension or pure shear deformation.

$$G_{IC}+\left(G_{IIC}\text{-}G_{IC}\right)\left(\frac{G_{II}}{G_T}\right)^{\eta}=G_{TC} \tag{2}$$

In the formula, G_I, G_{II} and G_T are the energy release rate of type I, type II and the total energy release rate respectively. G_{IC}, G_{IIC} and G_{TC} are type I, type II critical energy release rate and composite critical energy release rate, η is a material parameter, which is taken as 1.45 here.The properties of the cohesive layer elements are shown in the table below:

Table 1. Property parameters of cohesive element material.[2]

ρ/ton·mm^{-9}	E_{nn}/GPa	E_{ss}/GPa	σ_n^0/MPa	$\sigma_s^0=\sigma_t$/MPa	G_{IC}/kJ·m^{-2}	$G_{IIC}=G_{IIIC}$/kJ·m^{-2}
1.65×10^{-9}	7.63	4.36	26	54	0.28	0.81

The plate adopts reduced integration, hourglass-controlled eight-node linear hexahedron element (C3D8R), and the cohesive layer bonding element adopts viscous element (COH3D8). For the damage propagation under each damage mode, the implicit solution, the explicit solution without mass scaling, and the explicit solution with the incremental step increasement of 2×10^{-6} as the mass scaling target are used for analysis. And compare the evolution of the support reaction with the loading displacement at the reference point of the applied load.

According to the comparison of the results of the explicit solution and the implicit solution under the three different modes of damage, it can be seen that the ABAQUS/Explicit solver can better fit the calculation of the ABAQUS/Standard solver for the two types of damage expansion of damage mode I and mode III, for the extension of damage mode II, the solution effect of the explicit algorithm is not ideal. In order to improve the calculation efficiency, mass scaling can be set during the explicit solution process. In the explicit calculation of damage expansion of mode I and mode III, the explicit calculation with mass scaling is more effective than the explicit calculation without mass scaling. The error of the explicit calculation with mass scaling is within 1% compared with the explicit calculation without mass scaling. Therefore, for this two modes of damage, the explicit calculation with mass scaling can be adopted.

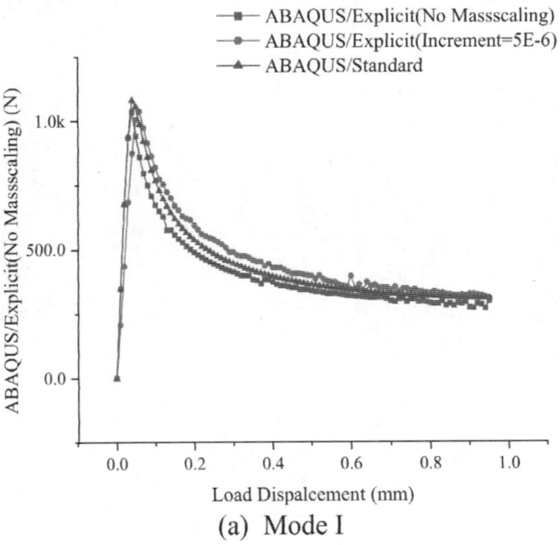

(a) Mode I

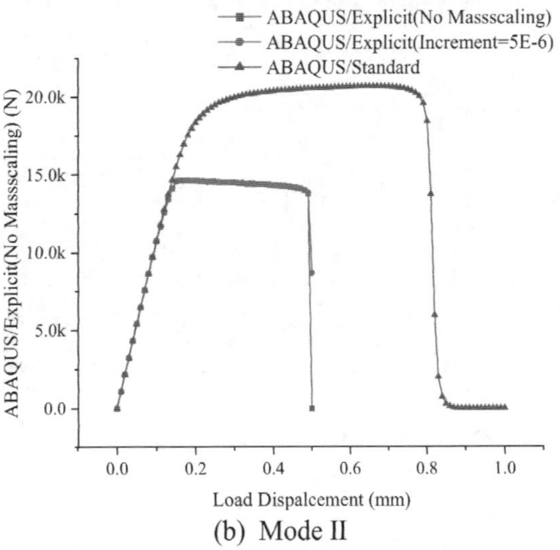

(b) Mode II

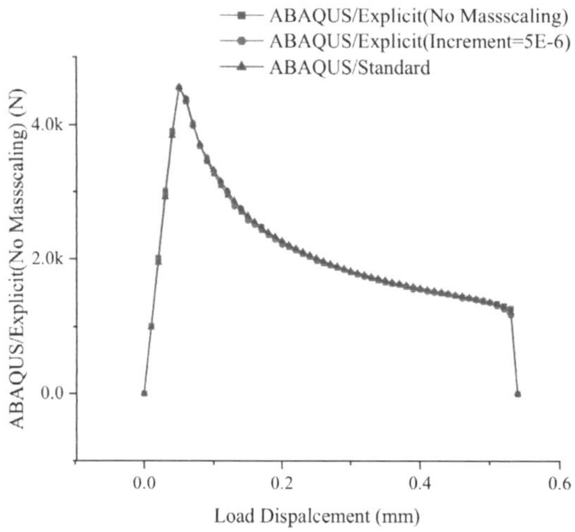

(c) Mode III

Figure 2. Variation of compressive force with loading under different modes of damage.

3. Analysis of compressive strength of composite laminates under different preset delamination damage ranges

As mentioned above, in this example, the ABAQUS/Explicit solver is used to simulate the delamination expansion of a composite laminate with internal preset delamination damage, the compression performance of the composite laminate under different levels of preset delamination is explored. According to the ASTM-D7136 test standard[12], the length and width of the composite laminate used in this example is 150mm×100mm, with a total of 11 layers of T800 carbon fiber reinforced epoxy resin-based unidirectional tape prepreg layup, the length direction of the plate is 0° direction, the layup is set to $[0°/90°/0°/90°/0°/\overline{90}°]_s$, each layer of unidirectional prepreg is 0.2mm, insert cohesive elements between the third and the forth layers. The mechanical parameters of ply in the laminate are as table 2:

Table 2. Mechanical parameters of unidirectional prepreg of T800 CFRP.

E_1/MPa	E_2/MPa	E_3/MPa	Nu_{12}	Nu_{13}	Nu_{23}	G_{12}/MPa	G_{13}/MPa	G_{23}/MPa
173000	8500	8500	0.35	0.35	0.4	4500	4200	4200
X_t/MPa	X_c/MPa	Y_t/MPa	Y_c/MPa	Sc_{12}/MPa	Sc_{31}/MPa	Sc_{23}/MPa	ρ/ton·mm^{-9}	
2500	1500	64	260	110	110	79.9	1.65×10^{-9}	

The Hashin criterion is used to write a subroutine to analyze the damage of the fiber and the matrix during the impact and buckling of composite laminates. The criterions of different types of failure are as follows:

Table 3. Simplified Hashin failure criterion.[11]

Failure Type		Failure Criterion	
Fiber	Tensile	$\left(\frac{\sigma_{11}}{X_T}\right)^2+\alpha\left(\frac{\sigma_{12}^2+\sigma_{13}^2}{S_{12}^2}\right)\geq 1$	(3)
	Compress	$\left(\frac{\sigma_{11}}{X_C}\right)^2\geq 1$	(4)
Matrix	Tensile	$\left(\frac{\sigma_{22}+\sigma_{33}}{Y_T}\right)^2+\frac{\sigma_{23}^2-\sigma_{22}\sigma_{33}}{S_{23}^2}+\frac{\sigma_{12}^2+\sigma_{13}^2}{S_{12}^2}\geq 1$	(5)
	Compress	$\left(\frac{\sigma_{22}+\sigma_{33}}{2S_{23}}\right)^2+\frac{\sigma_{22}+\sigma_{33}}{Y_C}\left[\left(\frac{Y_C}{2S_{23}}\right)^2-1\right]+\frac{\sigma_{23}^2-\sigma_{22}\sigma_{33}}{S_{23}^2}+\frac{\sigma_{12}^2+\sigma_{13}^2}{S_{12}^2}\geq 1$	(6)

The composite laminates is fixed at one end along the length ($U_1=U_2=U_3=UR_1=UR_2=UR_3=0$), and the other end is subjected to a compressive displacement at the reference point and acts on the end face. 1% of the displacement of the first-order buckling mode is introduced as the initial damage, after setting the variable output, submit the job and post-process.
In order to make the nodes in a steady state before and after the incremental step as much as possible, the incremental time step should be small enough, the relevant physical quantities in each incremental step are approximately constant. The stable incremental time step is as follow:

$$\Delta t_{stable} = \frac{L_e}{C_d} \tag{7}$$

In the formula, L_e is the characteristic length of the element, generally adopts the shortest element size, C_d is the wave velocity of the material. For a linear elastic material with Poisson's ratio zero:

$$C_d = \left(\frac{E}{\rho}\right)^{\frac{1}{2}} \tag{8}$$

In the formula, E is the elastic modulus of the material, ρ is the density of the material. During the calculation process, the stability time increment of the analysis step is about 6×10^{-7}.

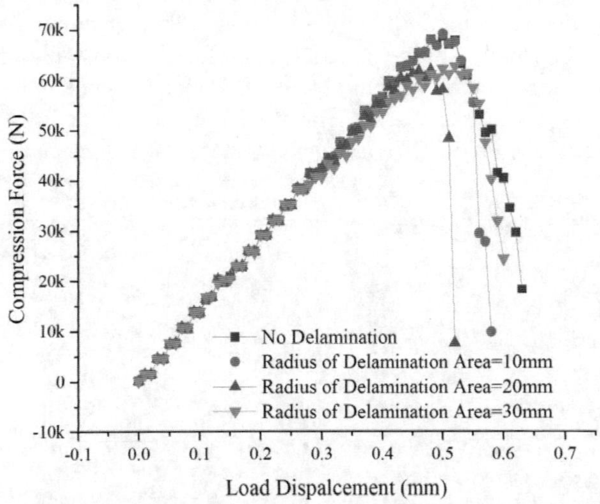

Figure 3. Variation of compressive force with loading under different preset damages.

The compression force of composite laminates under different pre-delaminated areas varies with loading conditions is shown in the figure. It can be seen from the table that for a laminate without pre-delaminated and with a pre-delaminated circle area of 10 mm, there is no significant difference in the compressive strength, about 70kN, but for the laminate with the pre-delaminated circle area of 20mm and 30mm, the compressive strength is significantly reduced, about 60kN, the larger the damage area, the faster the structural damage. Therefore, in actual engineering, it can be considered that a circle with a delamination area below 10mm is acceptable.

4. Simulation analysis of compressive strength of single-fastener connected composite laminates with pre-delaminated

In this example, the standard single-fastener connected test piece in ASTM D5961M-08[13] is used as the simulation object, two composite plates with ply scheme $[0°/90°/0°/90°0°/90°/0°/\overline{90°}]_S$ are connected by countersunk fasteners, and thickened reinforcement areas are arranged at both ends. Clamps are used to clamp and apply loads in the test. In this example, delamination is arranged on the inside of each composite plate at 1/4 thickness from the outer surface, the pre-delamination area is a circle with a radius of 12mm, which is concentric with the connection hole.

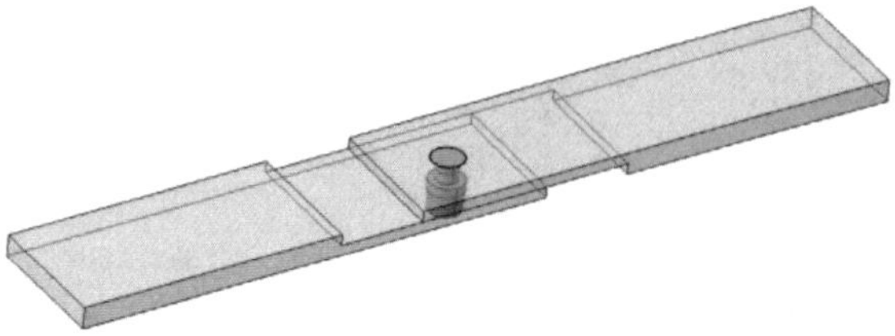

Figure 4. Standard test piece for bearing response of polymer matrix composite laminates.

In this example, the buckling mode of the connected composite structure is first calculated, 1% of the first-order mode displacement is extracted as the initial damage. One end of the plate connector is fixed, the other end is fixed, both the sides limit the out-of-plane displacement. The self-written subroutine based on the three-dimensional Hashin failure criterion is used to analyze the fiber damage, matrix damage and delamination expansion during the compression process.

Through finite element simulation analysis, it can be obtained that when the analysis step time reaches 0.2, the composite material in the area around the connecting hole begins to be damaged due to compression; when it reaches 0.34, the damage trend begins to become obvious; when it reaches 0.8, the damage of fibers and matrix has already becomes serious, at which point the preset delamination begins to expand. It can be seen from the numerical simulation that for this kind of connection structure, due to the effect of the connecting fasteners, the delamination damage area expansion in the local area near the connecting area does not play a dominant role in the failure mode, the interlaminar failure of the material at this time is caused by the intralaminar failure of the material. Therefore, for this kind of small-scale preset delamination damage, the failure caused by structural strength damage can be more considered when designing the structure.

Table 4. Destruction of connected regions at different analysis steps.

Step Time	Fiber destruction	Matrix damage	Interlayer damage
T=0.25		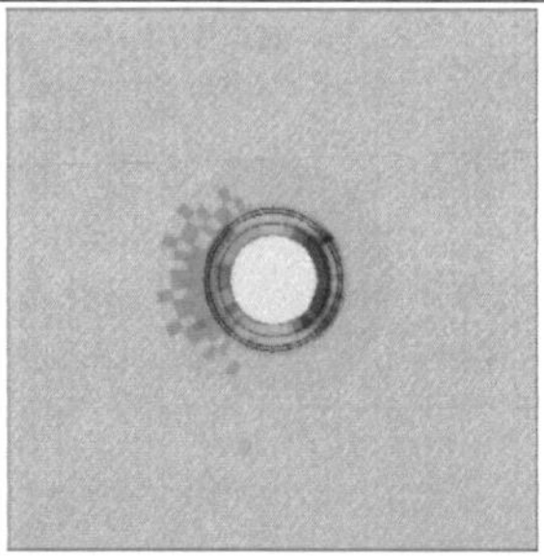	
T=0.5	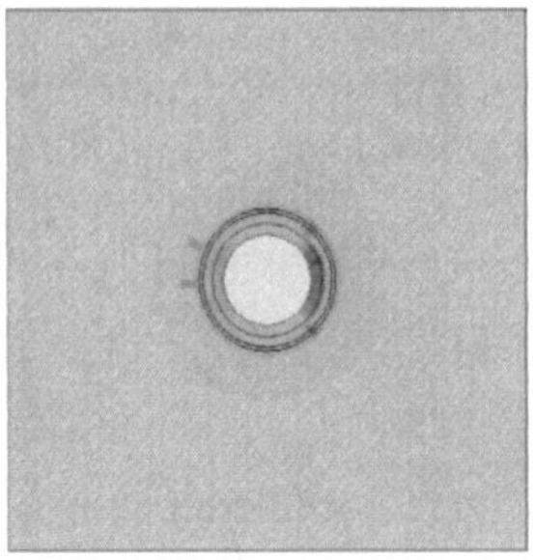	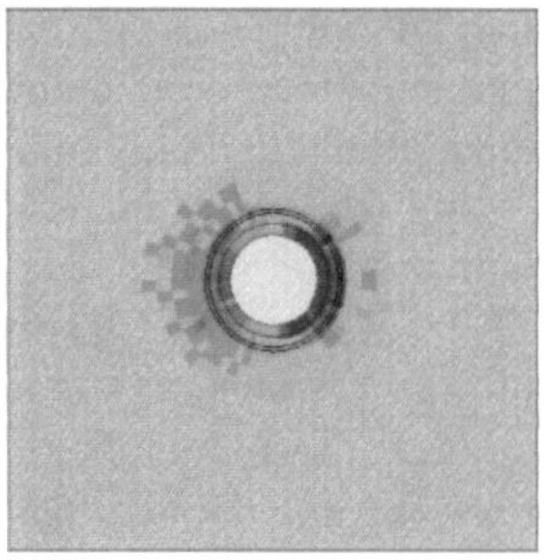	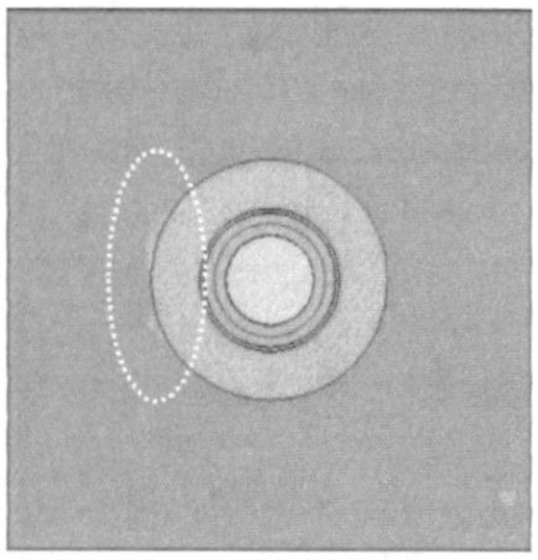
T=0.9	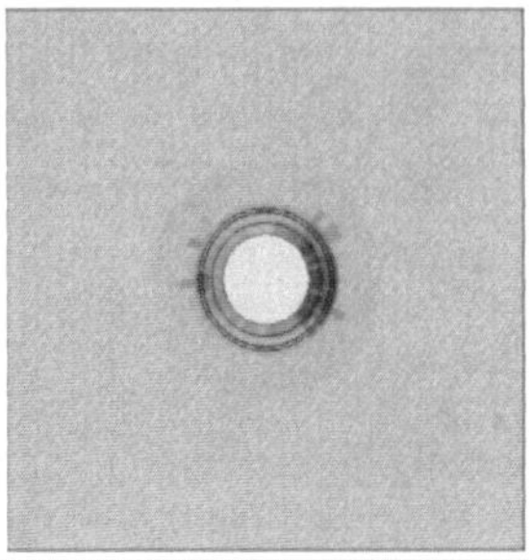	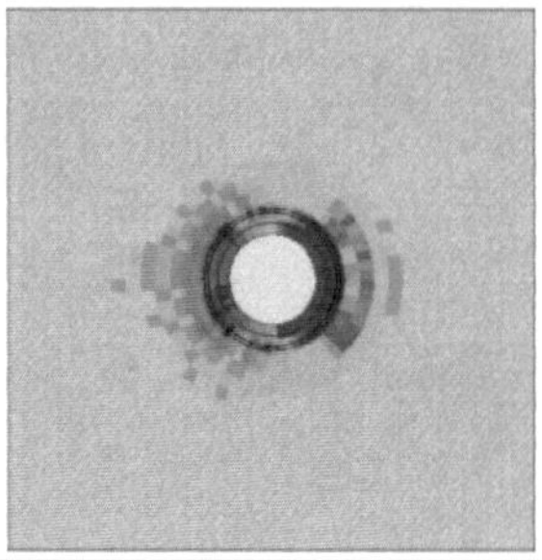	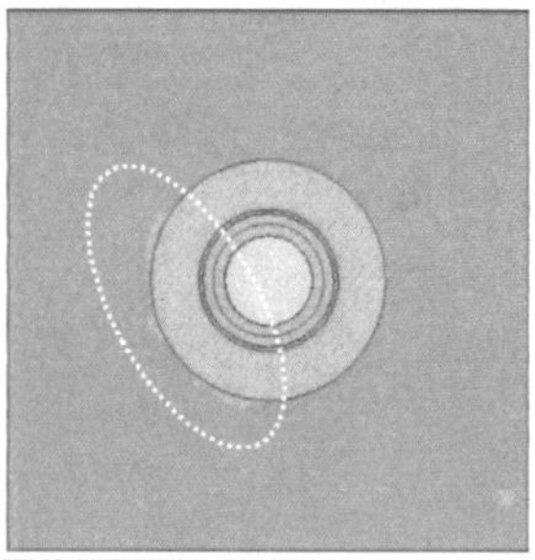

5. Conclusion

1. The ABAQUS/Explicit solver can be used to solve the delamination expansion simulation of the damage mode I and damage mode III. On the premise of meeting engineering accuracy, the use of explicit solvers in these two types of damage modes can effectively solve the problem of difficulty in convergence of implicit solvers in interlayer damage expansion analysis, and significantly improve the calculation speed.
2. For this kind of T800 carbon fiber reinforced epoxy resin-based unidirectional composite laminate with prepreg layup, the delamination damage below 10mm will not cause significant damage to the static compressive strength. For the delamination damage in tiny areas that cannot be detected by the existing ultrasonic C-scan detection level during the molding and use of composite materials, it can be considered that it will not cause destructive effects on the structural strength.
3. For this type of single-fastened composite laminates with pre-delamination, since the delamination area is very small, the pretightening force of the fastener connection is more significant at this time, which limits the deflection of the material in the local area under pressure deformation, the local delamination damage around the fastener does not dominate the damage failure around the connection structure.

References

[1] Fu C 2023 Research on delamination damage mechanism and prediction method of fiber reinforced composite laminates [D]. Beijing Jiaotong University,. DOI: 10.26944/d.cnki.gbfju.2022.000022.

[2] Zhang M H, Zhou C W, Lu H 2019 Numerical simulation of drilling delamination damage of carbon fiber composite unidirectional plate [J]. *Mechanical Engineering Materials*, **43**(09):73-77.

[3] Gu Z P 2016 Research on Delamination Failure Mechanism of Composite Materials Based on Cohesion Theory[D]. Zhejiang University.

[4] Xiao M L, Zhang Y B, Wang Z H, Fu H M and Yang X 2016 Effect of delamination damage on residual strength of composite laminates with holes[J]. *Journal of Aerodynamics*, **31**(05): 1081-1086. DOI: 10.13224/j.cnki .jasp.2016.05.008.

[5] Jia Z J 2020 Research on Residual Strength of Composite Laminates with Preset Damage [D]. Civil Aviation University of China, DOI: 10.27627/d.cnki.gzmhy.2018.000010.

[6] Wu W Y 2021 Numerical Simulation of Delamination Failure Behavior of CFRP Laminated Structure [D]. Beijing Institute of Technology,. DOI: 10.26948/d.cnki.gbjlu.2018.000586.

[7] Zhang L 2014 Delamination expansion behavior and numerical simulation of composite laminates with delamination defects[D]. Harbin Institute of Technology.

[8] Zhao L B, Gong Y, Zhang J Y, Chen Y L and Fei B J 2014 Simulation of Multidirectional Laminate Type I and Type I/II Delamination Expansion Based on Cohesion Model (English) [C]//Chinese Society of Mechanics. The 15th North Seven Provinces Proceedings of Urban Mechanics Academic Conference. Zhengzhou University Press, **vol 1**.

[9] Maruschak P O, Panin S V, Stachowicz F, Danyliuk I M and Vlasov I V 2016 Structural levels of fatigue failure and damage estimation in 17Mn1Si steel on the basis of a multilevel approach of physical mesomechanics[J]. *Acta Mech*, **227**, 151–157.

[10] Tan R M 2020 Study on the Issues of Low-Velocity Impact Damage Tolerance of Composite Structures [D]. Beihang University.

[11] Abaqus. 2014 ABAQUS user's manual. In Version 6.14 User's Guide, Abaqus: Providence, RI, USA; p. Section 10.7.1.

[12] ASTM D7136/D7136M 2007 Standard Test Method for Measuring the Damage Resistance of a Fiber-Reinforced Polymer Matrix Composite to a Drop-Weight Impact Event, ASTM International.

[13] ASTM D5961M-08 2008 Standard test method for Bearing response of polymer matrix composite laminates, ASTM International.

Marangoni convection in a C-shape enclosure with partially heated walls

Zailan Siri*, **Sharifah Nuriza Syed Muhammas Naquid Al 'Idrus**

Institute of Mathematical Sciences, Universiti Malaya, 50603 Kuala Lumpur, Malaysia

* Corresponding author's email: zailansiri@um.edu.my

Abstract. A numerical study was carried out to investigate Marangoni convection of nanofluid in a C-shape cavity with partially heated walls. The opposite sides of the walls are cooled at constant temperature while the rest of the partitions are kept adiabatic. The governing equations and boundary conditions are then introduced to describe the fluid flow and temperature distribution within the enclosure before the equations are non-dimensionalised and solved using the finite element method. The solutions, presented as streamlines, isotherms, local Nusselt and average Nusselt for varying Marangoni number, Rayleigh number, and depth, are then discussed.

1. Introduction

With the ever-increasing dependence on technology and machines in most, if not all industries, the demand for better cooling solutions and fluids that are capable of transferring heat at a high rate has never been greater. One such option is the Marangoni convection (*Ma*), which has been a topic of interest for the past few decades. Some of its applications include electronic parts cooling, solar water heating and atmospheric fallout. Several investigations and experiments have been conducted that considered the surface tension temperature gradient under various conditions and cavities, including C-shape enclosures. Mansour *et al.* [1] investigated natural convection fluid circulation and heat transfer inside a C-shape enclosure filled with nanofluid. It was discovered that the heat transfer rate improved as the aspect ratio of the cavity decreased. Mohebbi *et al.* [2] also conducted a study of natural convection in a C-shape enclosure, though the cavity contained a heat source that was considered in various locations inside the enclosure. They concluded that when the Rayleigh number (*Ra*) is high and that when the heat source was in a vertical wall, the Nusselt numbers (*Nu*) achieved its highest values.

Haq *et al.* [3] investigated heat transfer inside a C-shape cavity with partially heated walls filled with Single Wall Carbon Nanotubes. It was found that increasing the *Ra* helped enhance the stream flow and isotherms behaviour, though increasing the Hartmann number (*Ha*) had the opposite effect. Bakier [4] studied natural convection in a partially heated C-shape enclosure with an open end. The cavity was filled with nanofluid. He concluded that the presence of nanoparticles (ϕ) helped improve heat and mass transfer through the open section when the *Ra* is small. In addition to that, the temperature of the cavity increased as the heat source length became longer.

Nia *et al.* [5] investigated natural convection heat transfer of nanofluid in an L-shape enclosure with curved boundaries. They concluded that the curved L-shape models increased the *Nu* values significantly. Lugarini *et al.* [6] numerically studied natural convection and surface radiation in a heated wall with a C-shape fracture. It was concluded that heat transfer decreased when radiation was present. Yildiz *et al.* [7] investigated natural convection of nanofluid in a U-shape enclosure with cold ribs where the inside walls of the cavity are cooled. It was found that the average Nusselt number (Nu_{av}) increased with the *Ra*, though this becomes insignificant as the aspect ratio of the enclosure gets higher.

The purpose of this study is to investigate *Ma* of nanofluid in a C-shape enclosure with two heated walls and three cooled walls. *Ma* occurs at the top of the cavity and the depth (*d*) is varied for *Ma*, *Ra* and ϕ.

2. Mathematical formulation

The two-dimensional representation of *Ma* inside a nanofluid C-shape enclosure as shown in Figure 1 was considered with height (*H*) and length (*L*) where in this study both are assumed to be equal. The distance AD is denoted as *d*, while *a* is the length of BC and GF. The heated walls are denoted by T_h while the cold walls, T_c. Marangoni convection transpires at the top wall GH while the rest of the partitions are adiabatic (BC and FG).

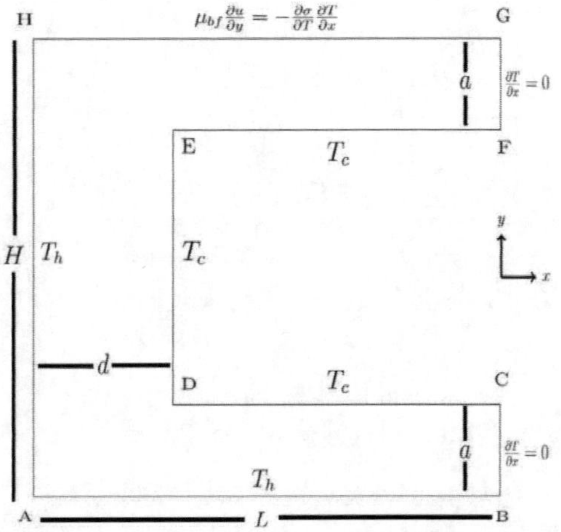

Figure 1. Two-dimensional physical model of the C-shaped enclosure

The nanofluid inside the cavity is water-based nanofluid with Al_2O_3 nanoparticles (H_2O-Al_2O_3). The thermo-physical properties are shown in Table 1.

Table 1. Thermo-physical properties of H_2O and Al_2O_3

Physical Property	H_2O	Al_2O_3
$\rho[kg \cdot m^{-3}]$	997.1	3600
$C_p[J \cdot (kg) - 1 \cdot K^{-1}]$	4179	765
$k[W \cdot m^{-1} \cdot K^{-1}]$	0.605	46
$\beta \times 10^{-5}[K^{-1}]$	21	0.63

For this study fluid flow is laminar, steady, and incompressible with viscous dissipation being absent. This assumption is made by assuming the fluid flows in infinitesimal parallel layers with no disruption between them. By using the following substitutions,

$$X=\frac{x}{L},\ Y=\frac{y}{L},\ U=\frac{uL}{\alpha_f},\ V=\frac{vL}{\alpha_f},\ \theta=\frac{T-T_c}{T-T_h},\ P=\frac{pL^2}{\rho_{nf}\alpha_f^2},$$

$$\Pr=\frac{\mu_f}{\rho_f\alpha_f},\ Ra=g\beta_f L^3\frac{T_h-T_c}{\nu_f\alpha_f},\ Ma=-L\frac{\partial\sigma}{\partial T}\frac{T_h-T_c}{\mu_f\alpha_f},\ \Psi=\frac{\psi}{\alpha_f}.$$

The governing equations and boundary conditions based on these assumptions are,

$$\frac{\partial U}{\partial X}+\frac{\partial V}{\partial Y}=0, \tag{1}$$

$$U\frac{\partial U}{\partial X}+V\frac{\partial U}{\partial Y}=-\frac{\partial P}{\partial X}+\Pr\frac{\mu_{nf}/\mu_f}{\rho_{nf}/\rho_f}\left(\frac{\partial^2 U}{\partial X^2}+\frac{\partial^2 U}{\partial Y^2}\right), \tag{2}$$

$$U\frac{\partial V}{\partial X}+V\frac{\partial V}{\partial Y}=-\frac{\partial P}{\partial Y}+\Pr\frac{\mu_{nf}/\mu_f}{\rho_{nf}/\rho_f}\left(\frac{\partial^2 V}{\partial X^2}+\frac{\partial^2 V}{\partial Y^2}\right)+\frac{\beta_{nf}}{\beta_f}Ra\Pr\theta, \tag{3}$$

$$U\frac{\partial \theta}{\partial X}+V\frac{\partial \theta}{\partial Y}=\frac{\alpha_{nf}}{\alpha_f}\left(\frac{\partial^2 \theta}{\partial X^2}+\frac{\partial^2 \theta}{\partial Y^2}\right), \tag{4}$$

$$\frac{\partial V}{\partial X}-\frac{\partial V}{\partial Y}=-\nabla^2\Psi, \tag{5}$$

with boundary conditions,

$$\left.\begin{array}{ll}
U=V=0, & \text{for walls AB, BC, CD, DE, EF and HA,}\\
\dfrac{\partial\theta}{\partial X}=0, & \text{for walls CB and GF,}\\
\dfrac{\partial\theta}{\partial Y}=0, & \text{for wall GH,}\\
\theta=1, & \text{for walls HA and AB,}\\
\theta=0, & \text{for walls CD, DE and EF,}\\
\dfrac{\partial U}{\partial Y}=\mathrm{Ma}\dfrac{\partial\theta}{\partial X}=0, & \text{for wall GH,}
\end{array}\right\} \tag{6}$$

The definition of other parameters used in this investigation are depicted in Table 2, While, the Nu_{av} is then computed by taking the integral of,

$$Nu_{av}=\int_0^1 \mathrm{Nu}_{\mathrm{loc}}\frac{\partial\theta}{\partial N}dN. \tag{7}$$

Table 2. Parameters definition

Definition	Parameter
nanoparticles density	$\rho_{nf} = (1-\phi)\rho_f + \phi\rho_s,$
heat capacitance	$(\rho C_p)_{nf} = (1-\phi)(\rho C_p)_f + \phi(\rho C_p)_s,$
dynamic viscosity	$\mu_{nf} = \dfrac{\mu_f}{(1-\phi)^{2.5}},$
thermal diffusivity	$\alpha_{nf} = \dfrac{\kappa_{nf}}{(\rho C_p)_{nf}},$
thermal expansion coefficient	$\beta_{nf} = (1-\phi)\beta_f + \phi\beta_s$
thermal conductivity	$\kappa_{nf} = \dfrac{\kappa_s + 2\kappa_f - 2\phi(\kappa_f - \kappa_s)}{\kappa_s + 2\kappa_f - \phi(\kappa_f - \kappa_s)},$
local Nusselt at the vertical heated walls	$Nu_{\text{loc}}^{v} = -\dfrac{\kappa_{nf}}{\kappa_f}\dfrac{\partial\theta}{\partial X}$
local Nusselt at the horizontal heated walls	$Nu_{\text{loc}}^{h} = -\dfrac{\kappa_{nf}}{\kappa_f}\dfrac{\partial\theta}{\partial Y}$

3. Results and discussion

In this investigation, COMSOL 5.3a was utilised to solve the governing equations (1) – (5) with respect to its boundary conditions (6) incorporated with the Finite Element Method (FEM) explained in Taylor and Hood [8] and Dechaumphai [9]. The current study was investigated numerically for the values of $0.3 \le d \le 0.7$, $0 \le Ma \le 10^3$ and $10^3 \le Ra \le 10^4$ with fixed Prandtl number, $Pr = 0.052$ and $\phi = 0.03$. The solutions obtained are compared with Saleh and Hashim [10] with the absolute error is less than 10% as tabulated in Table 3.

Table 3. Comparisons of Nu_{av} for $Ra = 10^3$, $Pr = 0.052$ and $\phi = 0$

Ma	**Saleh and Hashim [10]**	**Present Study**
0	1.1492	1.1646
10^2	1.1681	1.0709
10^3	1.4507	1.5452

Figure 2 illustrate fluid flow inside the enclosure, when $Ma = 0$, a single cell circulates counter-clockwise inside the cavity and the strength of flow increases as d increases. When $Ma \neq 0$, the single cell splits into two with the top one rotating clockwise and the bottom, counterclockwise. Furthermore, the top cell circulates faster and gets bigger as d increases. For $Ra = 10^4$, the cells for all parameters oscillate at a higher rate and when $d = 0.7$, the cell retains its shape rather than splitting itself up. Increasing *Ma*, its produces two separate cells with the bottom one having a slightly weaker flow for the value of $d = 0.3$, whereas the bottom cell for $d = 0.7$ oscillates faster compared to the one closer to where *Ma* takes place.

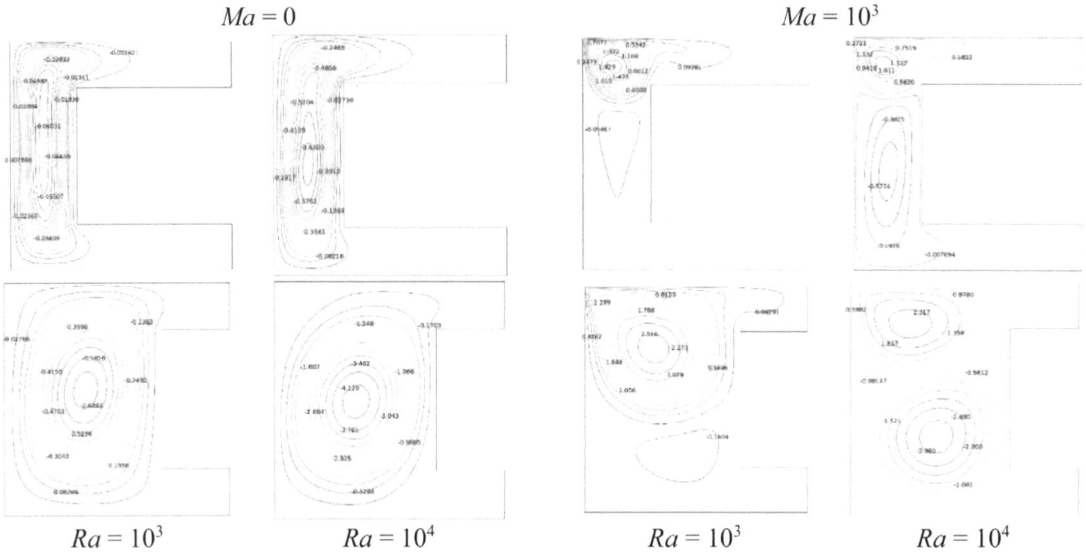

Figure 2. Streamlines for $d = 0.3$ (top) and $d = 0.7$ (bottom)

Figure 3 show the temperature distribution inside the enclosure. There is small to no difference in the contours when $Ma = 0$, the isotherms change drastically with increasing d and Ma. The most noticeable deformations are observed when $Ma = 10^3$ where temperature distribution at the top of the cavity curve into a boomerang like shape for $d = 0.3$, while for $d = 0.7$, the deformation occurs in the mid-section of the cavity. For the case when $Ra = 10^4$, the contours are much more prominent, even when $Ma = 0$.

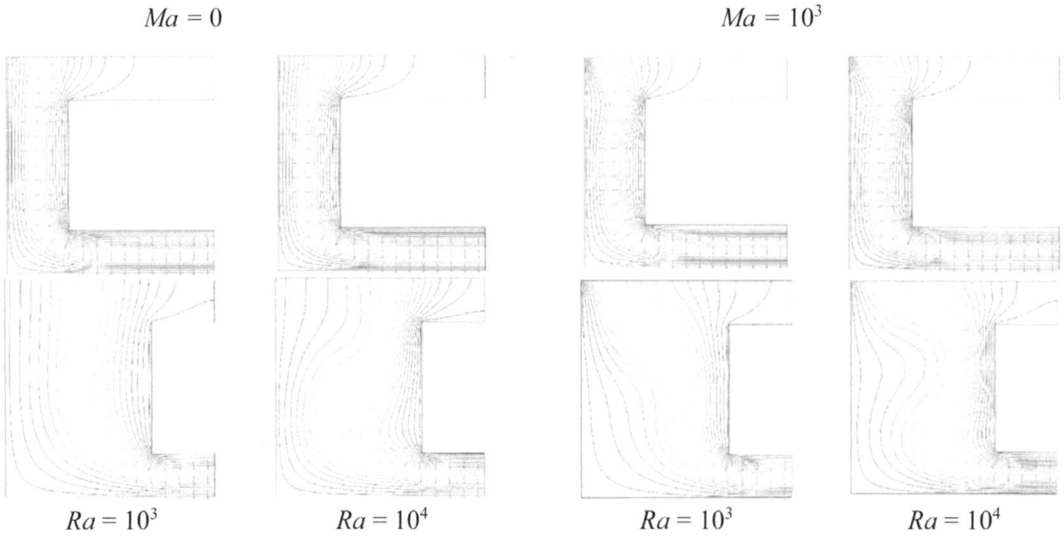

Figure 3. Isotherms for $d = 0.3$ (top) and $d = 0.7$ (bottom)

The Nu_{loc}^{v} and Nu_{loc}^{h} is depicted in Figure 4. The highest values of Nu_{loc}^{v} can be found in the middle of the enclosure for $Ma \leq 10^2$, that is $0.4 \leq Y \leq 0.5$, this is due to the cold wall being in the centre.

However, when $Ma = 10^3$, the highest values are located closer to $Y = 1$. As Ma becomes more intense, the surface tension gradient increases, causing the temperature gradient to decrease. Examining the horizontal wall, it can be said that for all parameters, the behaviour of the profiles is similar, with the highest values of Nu^h_{loc} achieved at $Y \geq 0.3$, this is due to the location of the wall itself where the cold wall is situated above the heated wall and the bottom wall being the farthest from where Marangoni convection takes place.

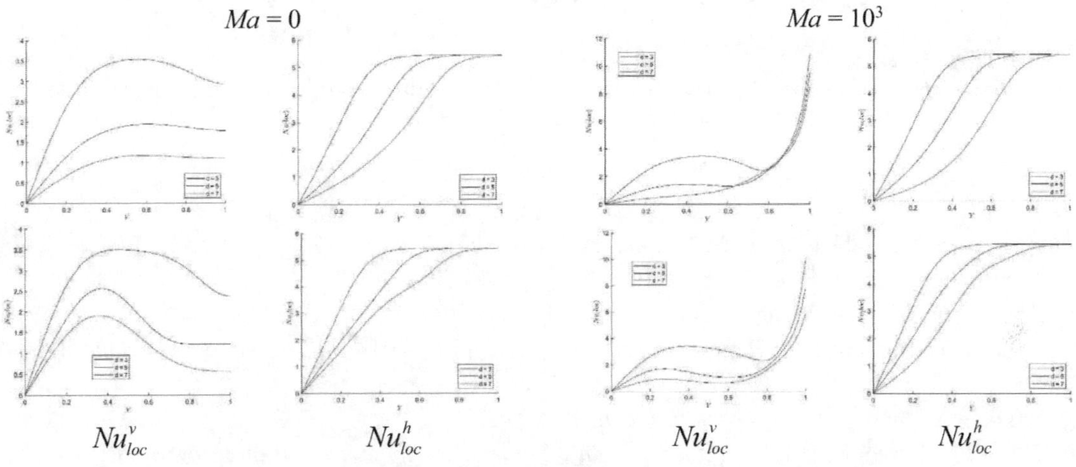

Figure 4. Nu^v_{loc} and Nu^h_{loc} for $Ra = 10^3$ (top) and $Ra = 10^4$ (bottom)

Generally, Nu_{av} for $T_x + T_y$, it increases steadily with Ma and decreases when d increases as shown in Fig. 5. Also, it can be stated that for all cases, Nu_{av} increases linearly for both cases when $Ma = 0$ and $Ma = 10^3$ where the highest heat transfer rates are obtained by $d = 0.3$.

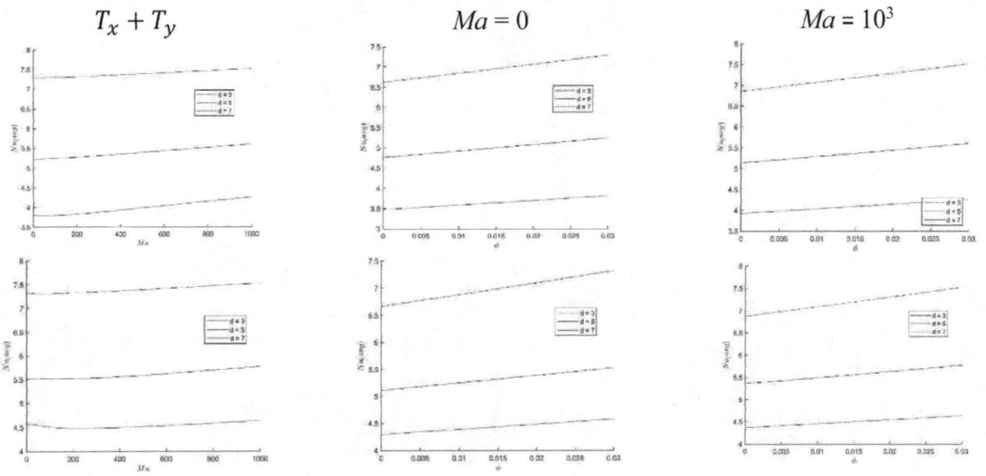

Figure 5. Nu_{av} of $T_x + T_y$ for $Ra = 10^3$ (top) and $Ra = 10^4$ (bottom)

4. Conclusion

Numerical study was conducted for *Ma* inside a C-shape cavity filled with nanofluid. The governing equations and boundary conditions based on the problem are non-dimensionalised and then solved using FEM before the results were collected and presented as streamlines, isotherms, local Nusselt and average Nusselt for various values of *Ma*, *Ra* and *d*. It can be concluded that,

- the strength of the flow increases with *Ma*, *Ra* and *d*,
- the highest values of Nu_{loc} and Nu_{av} were around $d = 0.3$ due to the hot and cold walls being closer, and
- *Ma* has no effect on the horizontal heated wall.

Acknowledgments

This work is supported by the Ministry of Higher Education Malaysia under the Fundamental Research Grant Scheme FRGS/1/2020/STG06/UM/02/4 and the Universiti Malaya for awarded Research Leave in 2023.

References

[1] Mansour M, Bakeir M, and Chamkha A J 2014 *Int. J. Numer. Methods Heat Fluid Flow* **24** p 1954
[2] Mohebbi R J, Izadi M, and Chamkha A J 2017 *Phys. Fluids* **29** p 122009
[3] Haq R U, Soomro F A, Hammouch Z, and Rehman S U 2018 *Int. J. Heat Mass Transf.* **127** p 506
[4] Bakier M 2014 *Eng. Sci. Technol. Int. J.* **17** p 116
[5] Nia S N, Rabiei F, and Rashidi M 2021 *Int. J. Numer. Methods Heat Fluid Flow* **32** p 1540
[6] Lugarini A, Franco A T, Junqueira S L, and Lage J L 2018 *J. Heat Transfer* **140(8)** p 082501
[7] Yildiz C, Arici M, Karabay H, and Bennacer R 2020 *J. Therm. Anal. Calorim.* **93** p 1
[8] Taylor C and Hood P 1973 *Comput. Fluids* **1** p 73
[9] Dechaumphai P 1999 Finite Element Method in Engineering, *Chulalongkorn University Press*
[10] Saleh H and Hashim I 2015 *Appl. Math. Mech.* **36** p 1169

Numerical simulation of rebar corrosion in track bed of shield tunnel under stray current

Shuo Yu [1,*], Huajun Sun [2], Zijuan Wang [1] and Changbao Liu [1]

[1] Powerchina Huadong Engineering Corporation Limited, Hangzhou 310000, China
[2] Zhejiang provincial development and reform commission, Hangzhou 310000, China

* Corresponding author's eamil: yu_s@hdec.com

Abstract. Track bed is the important support structure at the bottom of shield tunnel, during metro operation, the leakage of stray current will lead to corrosion of rebar in the track bed, which will affect the safe operation of metro. In this paper, the corrosion calculation model of track bed rebar under stray current is established, and the effects of stray current input mode and concrete resistivity on the corrosion rate of track bed rebar are studied. The result shows the corrosion rate of track bed rebar decreases nonlinearly with the increase of concrete resistivity, and the reduction gradient of corrosion rate in bilateral input mode is greater than that in unilateral input mode. With the decrease of concrete resistance, the corrosion area will cover the left half in unilateral input mode and cover both ends of track bed rebar in bilateral input mode. When the input point voltage increases by 40%, the corrosion current density in the middle of track bed will increase by about 61%.

1. Introduction

Shield tunnel is the main structural form of metro tunnel [1], the track bed is poured at the bottom to support the sleeper. During metro operation, due to the decline of the insulation performance of rail fasteners, stray current will flow from track bed to tunnel structure, result in rebar corrosion inside track bed. As the corrosion product of rebar is 4-8 times that of the original rebar [2], when the corrosion-induced expansion pressure exceed the concrete tensile strength, the cracks will penetrate the concrete of track bed, which will reduce the stiffness of track bed and affect safe operation of the metro tunnel.

Different calculation methods and model tests have been used to study the corrosion mechanism of rebars in underground structure [3-7]. Through several groups of model tests, Bertolini et al. [3] studied the corrosion rate of rebar in underground concrete under the action of a direct current (DC) and alternating current (AC). Dolara et al. [4] developed a numerical model for a simplified shield tunnel and determined the corrosion area of rebars exposed to stray current. Lei et al. [5] studied the critical time of rebar corrosion in a shield tunnel segment using a model test and provided a durability evaluation model for shield tunnel segments. Based on real engineering examples and genetic programming, Gao et al. [6] proposed a calculation method for predicting the service life of tunnel structures subject to chloride-induced corrosion. Li et al. [7] established a corrosion calculation model of tunnel segment contained only a single row of main bars, research shows the 2D rust layer is distributed in an eccentric circle along the circumferential direction.

In the aforementioned studies, most research object focus on the main structure of the tunnel. However, relevant research shows that stray current has a great influence on the corrosion of track bed rebar, and the distribution form of rebar in the model test were unlike those of real track bed [8]. Some studies [9] have shown that the corrosion characteristics are not only affected by the stray current magnitude, but also by the distribution form of rebar. Therefore, the existing model cannot accurately describe the corrosion characteristics of track bed rebar.

In our research, the details of shield tunnel and track bed are introduced, and the corrosion calculation model of track bed rebar under stray current is established. At last the effects of stray current input mode and concrete resistivity on the corrosion rate of track bed rebar are studied.

2. Details of shield tunnel and track bed

Each ring of shield tunnel consists of six pieces. It includes one key block (K), two adjacent blocks (B1, B2), and blocks (A1, A2, A3). The key block has a centre angle of 21.5°, the adjacent blocks have centre angles of 67.5°, and the standard blocks three standard have centre angles of 68°, the two adjacent blocks is connected by bending bolts, as shown in Fig 1. The mesh type in the model is tetrahedral mesh element, and each segment contains rebar inside, as shown in Fig 2. The diameter of segment rebar and track bed are 18mm and 16mm respectively, and the diameter of track bed reinforcement is 16mm, as shown in Fig 3. The distribution of rebar in track bed structure as shown in Fig 4.

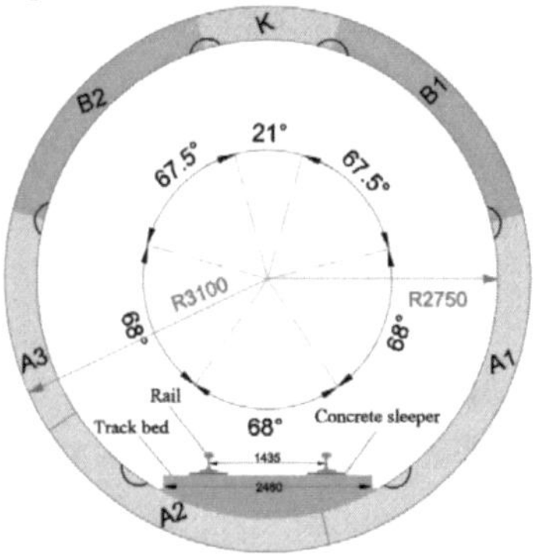

Fig. 1. Detail of shield tunnel

Fig. 2. Tetrahedral mesh model

Fig. 3. Rebar distribution in model

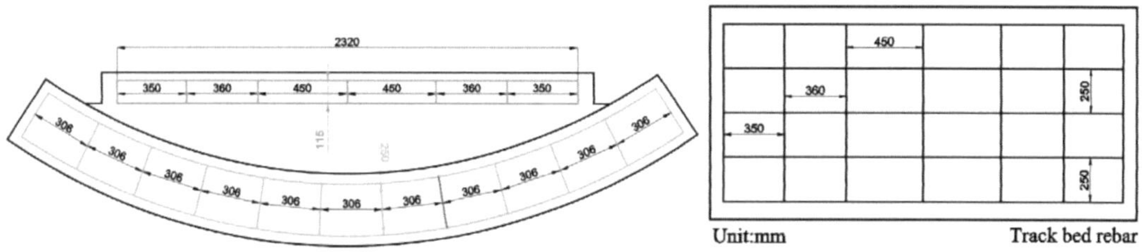

(a) Lateral view (b) Vertical view

Fig. 4. Size of track bed rebar

3. Corrosion calculation model of track bed rebar

According to Ohm's law [10], the current density vector $\vec{i}_T$ and $\vec{i}_S$ in concrete electrolyte can be expressed as:

$$\vec{i}_T = \sigma_T \vec{E}_T + \vec{i}_B \tag{1}$$

$$\vec{i}_S = \sigma_S \vec{E}_S + \vec{i}_B \tag{2}$$

Where the $\vec{i}_T$ and $\vec{i}_S$ are current density vector in track bed concrete and segment concrete respectively, the σ_T and σ_S are conductivity of track bed concrete and segment concrete respectively, the $\vec{E}_T$ and $\vec{E}_S$ are electric field intensity vector of track bed concrete and segment concrete respectively, the

$\vec{i}_B$ is the current density caused by external current, without considering the influence of other external currents, it is assumed that $\vec{i}_B$ is 0.

When the DC current flows through the tunnel structure and track bed, the generated potential difference will affect the electric potential E_e of the rebar surface. When the outside oxygen is sufficient, stray current will cause electrochemical corrosion to track bed rebar. The corrosion rate is closely related to the corrosion current density i_{corr}, which can be expressed according to the Butler-Volmer equation [11], as shown in Eq (3).

$$i_{corr}=i_{\mathrm{Fe},exc}\exp\left(2.303\frac{E_e-E_{\mathrm{Fe},eq}}{\beta_{\mathrm{Fe}}}\right) \tag{3}$$

Where i_{corr} is the corrosion current density; $i_{Fe,exc}$ is the anodic exchange current density, E_e is the electric potential; $E_{Fe,eq}$ is the anodic equilibrium potentials, β_{Fe} is the anodic Tafel slope. In the process of corrosion reaction, the corrosion rate is determined by the reaction area, as follows [12]:

$$R_{corr}=\frac{i_{corr}\cdot A_{Fe}}{m\cdot F\cdot\rho_{Fe}} \tag{4}$$

Where A_{Fe} is the mole mass of iron, ρ_{Fe} is the density of iron, the electrochemical calculation parameters as shown in Table 1. The software used for numerical calculation is COMSOL.

Table 1. Electrochemical calculation parameters [13].

Parameter	β_{Fe}	A_{Fe}	$E_{Fe,eq}$	ρ_{Fe}	$i_{Fe,exc}$
Value	0.09 (V/dec)	56 (g/mol)	-0.78 (V)	7.86 (g/cm^3)	3×10^{-4} (A/m^2)

During metro operation, due to the reduction of the rail fastener insulation performance, part of the stray current will flows from the rail to the track bed and segment structure, the current may flow from any point under the rail fastener. In the calculation model, it is assumed that the input location of stray current is the four points parallel on both sides of rail, selected point of X as the analysis object, as shown in Fig 5 (a) and (b). Relevant research shows that the stray current tent to flow out of the lower resistance part of metallic structure, In this paper the position where the two joints are assumed to be the current output boundary, as shown in Fig 5 (c).

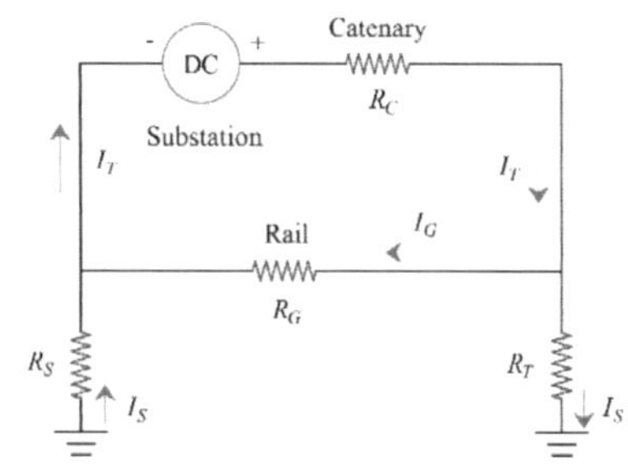

(a) Stray current leakage mode

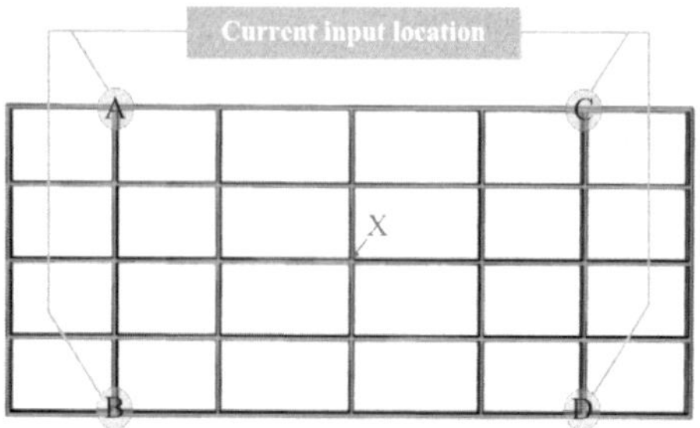

(b) Stray current input point

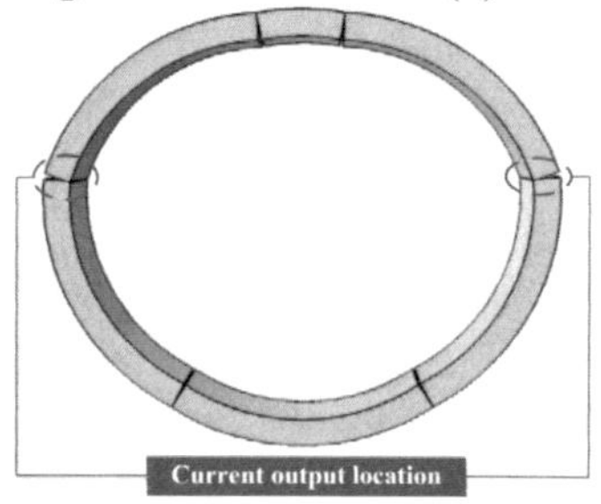

(c) Stray current output location

Fig. 5. Boundary conditions of calculation model.

4. Results and discussion

4.1. Unilateral input of stray current

Selected point A and B as the input location of stray current (unilateral input mode), each one point input voltage is 1V. Fig 6 shows the voltage distribution clound of track bed structure under different concrete resistivity when unilateral input, it can be seen that the smaller the concrete resistivity, the higher the voltage distribution near the rebar of track bed.

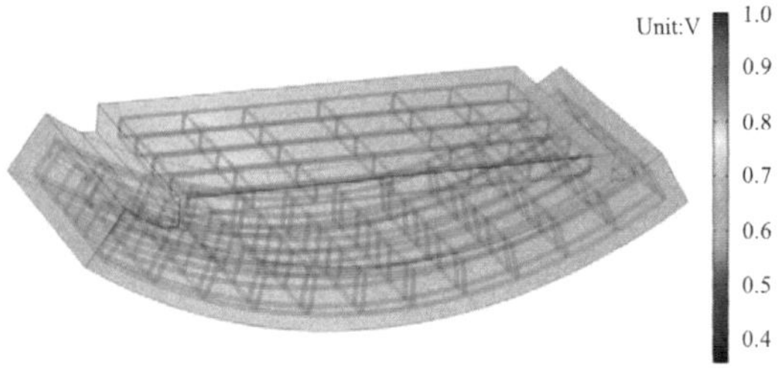

（a） Concrete resistivity is 1000Ω/m

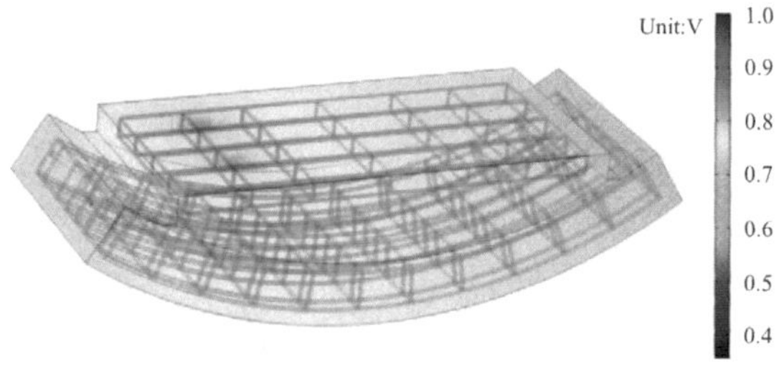

(b) Concrete resistivity is 800Ω/m

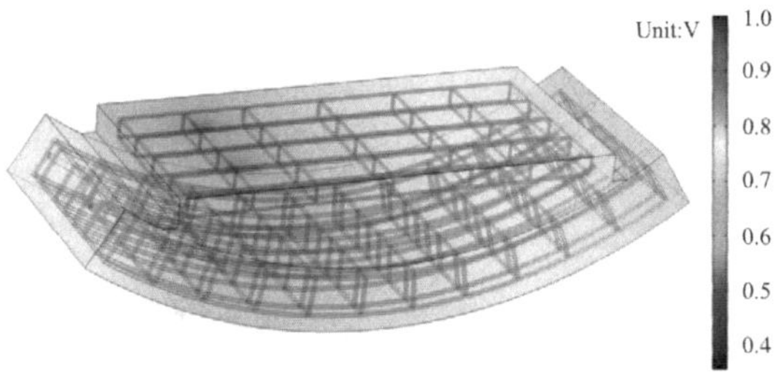

（c） Concrete resistivity is 600Ω/m

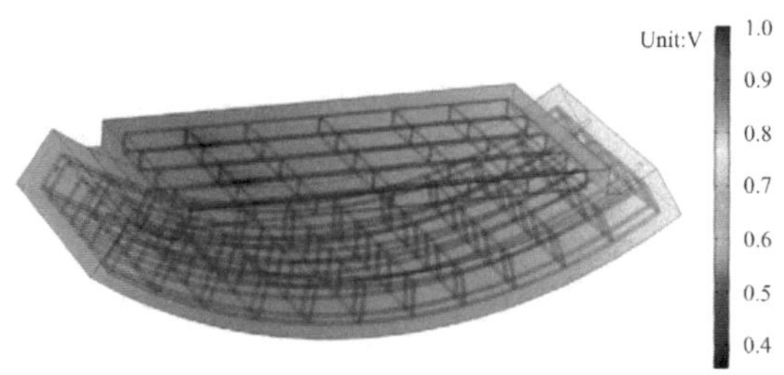

(d) Concrete resistivity is 400Ω/m

Fig. 6. Voltage distribution of track bed structure under different concrete resistivity when unilateral input.

Journal of Physics: Conference Series **2633** (2023) 012017 doi:10.1088/1742-6596/2633/1/012017

Fig 7 shows the corrosion current density distribution of track bed rebar under different concrete resistivity when unilateral input, it can be seen that when the concrete resistance is 1000 Ω/m, the corrosion area of track bed rebar concentrate near the input point. With the decrease of concrete resistance, the corrosion area extend from the upper to the lower rebar, and finally cover the left half of track bed rebar.

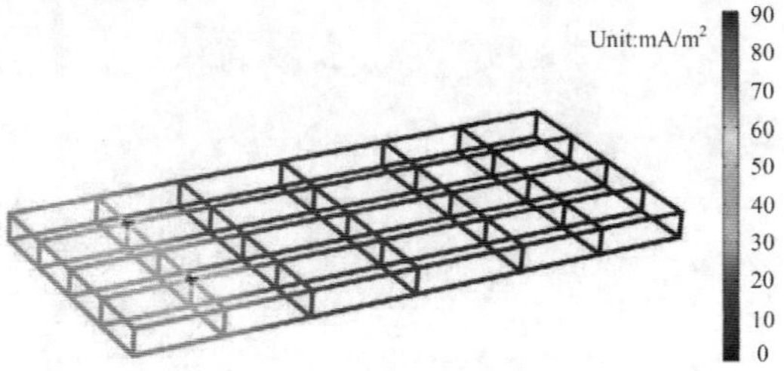

（a）Concrete resistivity is 1000Ω/m

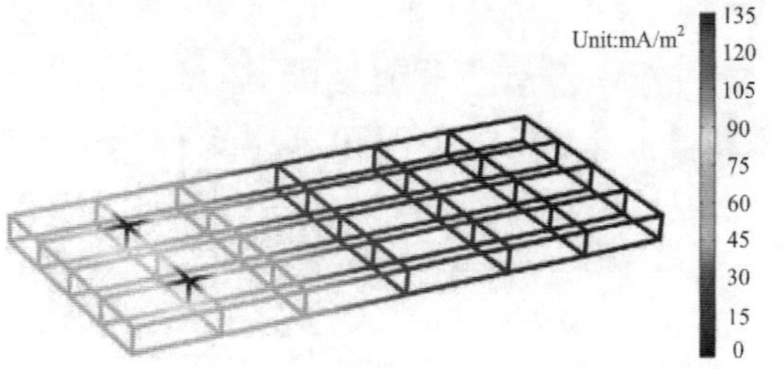

(b) Concrete resistivity is 800Ω/m

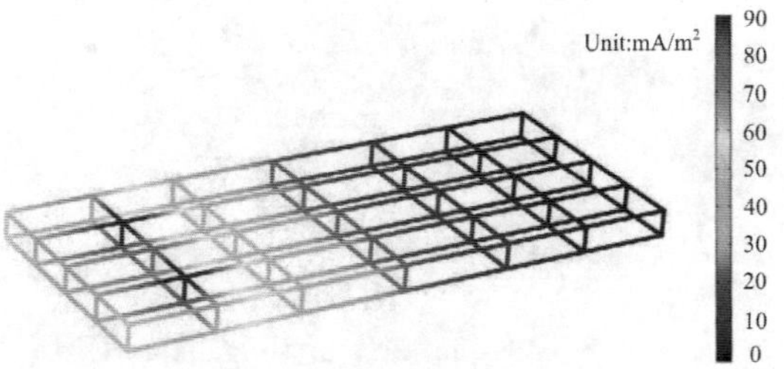

（c）Concrete resistivity is 600Ω/m

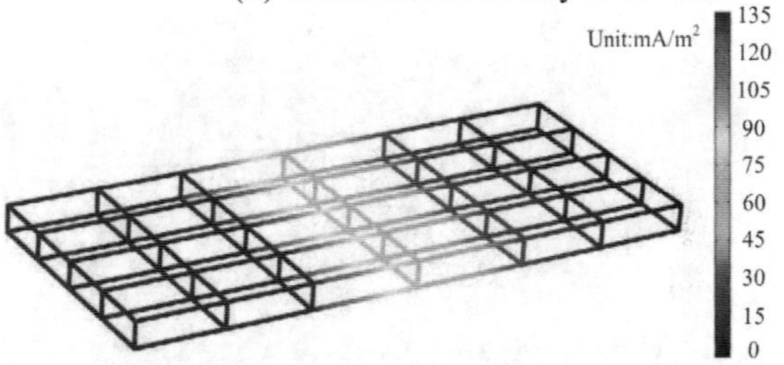

(d) Concrete resistivity is 400Ω/m

Fig. 7. Corrosion current density cloud picture of track bed rebar under different concrete resistivity when unilateral input.

Fig 8 shows the relation curve between point X voltage and input point voltage under different concrete resistivity, it can be seen that the voltage at point X increases linearly with the input point voltage, when the input point voltage increases by 40%, the voltage at point X will increase by about 28%. Fig 9 shows the relation curve between corrosion current density of point X and input point voltage under different concrete resistivity, it can be seen that the corrosion current density at point X increases nonlinearly with the input point voltage, when the input point voltage increases by 40%, the corrosion current density at point X will increase by about 52%.

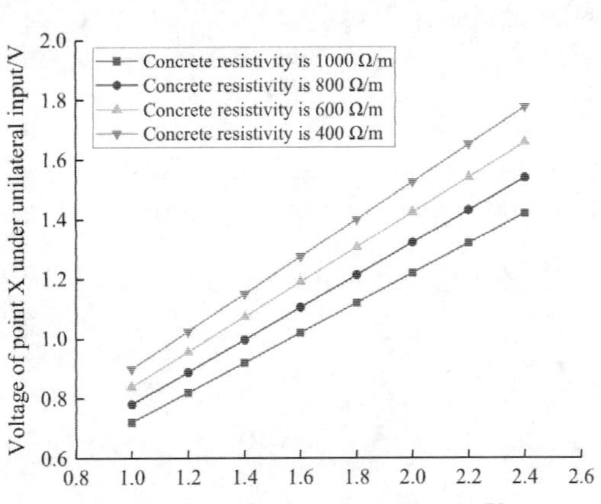

Fig. 8. Relation curve between point X voltage and input point voltage under unilateral input.

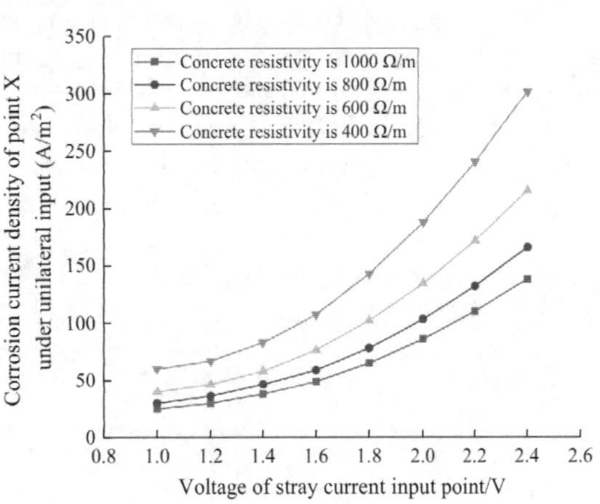

Fig. 9. Relation curve between corrosion current density of point X and input point voltage under unilateral input.

4.2. Bilateral input of stray current

Fig 10 shows the voltage distribution clound of track bed structure under different concrete resistivity when bilateral input, it can be seen that high voltage areas are concentrated in the middle, the voltage at both ends is small.

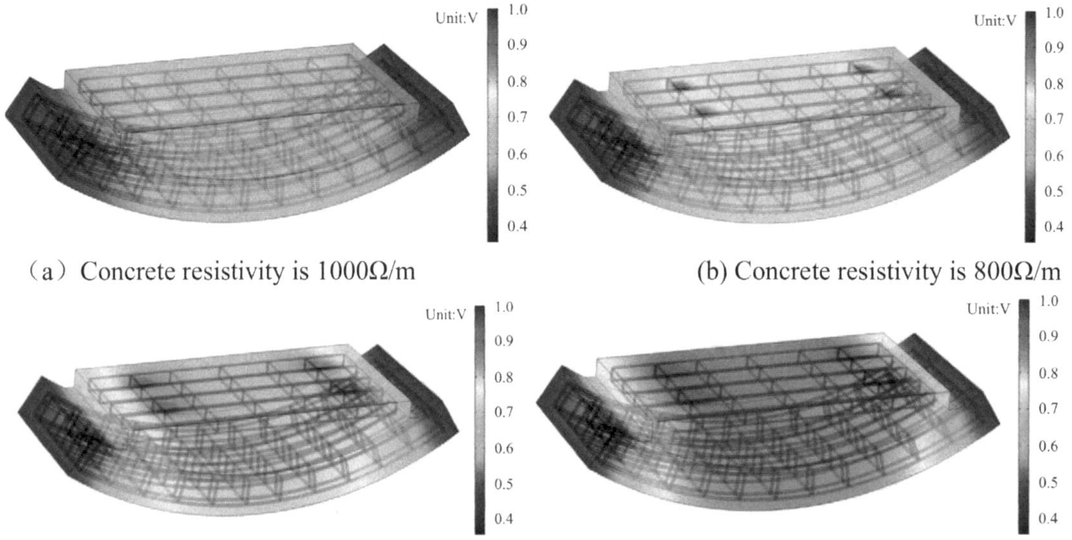

(a) Concrete resistivity is 1000Ω/m (b) Concrete resistivity is 800Ω/m

(c) Concrete resistivity is 600Ω/m (d) Concrete resistivity is 400Ω/m

Fig. 10. Voltage distribution of track bed structure under different concrete resistivity when bilateral input.

Fig 11 shows the corrosion current density distribution of track bed rebar under different concrete resistivity when bilateral input, it can be seen that when the concrete resistance is 1000 Ω/m, the corrosion area of track bed rebar concentrate near the upper rebar, with the decrease of concrete resistance, the corrosion area gradually expands downward, and finally cover both ends of track bed rebar.

Fig 12 shows the relation curve between point X voltage and input point voltage under different concrete resistivity at bilateral input, it can be seen that the voltage at point X increases linearly with the input point voltage, when the input point voltage increases by 40%, the voltage at point X will increase by about 35%. Fig 13 shows the relation curve between corrosion current density of point X and input point voltage under different concrete resistivity at bilateral input, it can be seen that the corrosion current density at point X increases nonlinearly with the input point voltage, when the input point voltage increases by 40%, the corrosion current density at point X will increase by about 61%.

Fig 14 and 15 shows the relation between input voltage, corrosion current density and concrete resistivity, the result shows under the bilateral input mode, the influence of concrete resistivity on rebar voltage and corrosion rate is more sensitive than that under the unilateral input mode.

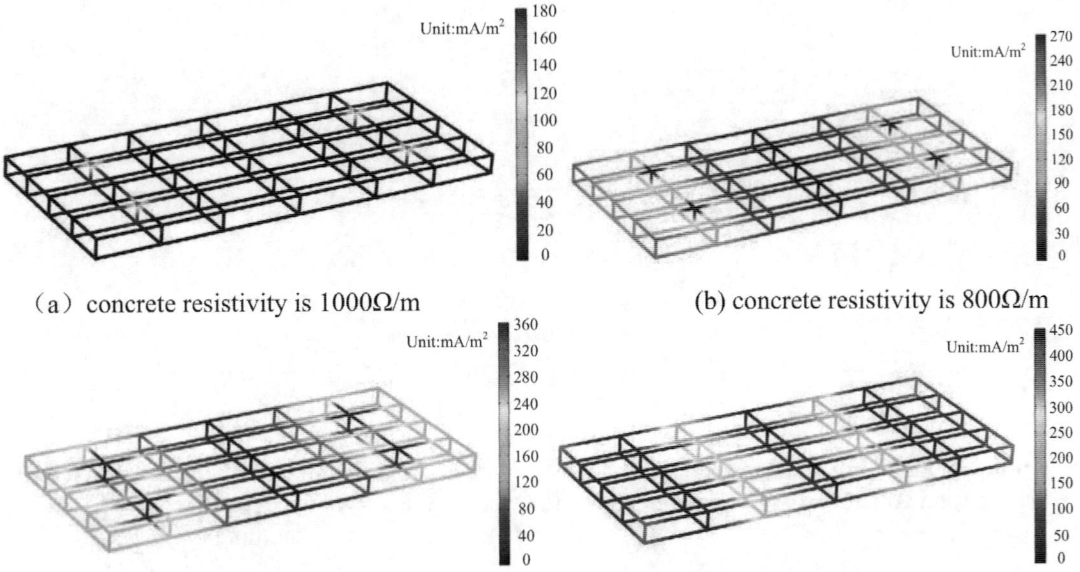

(a) concrete resistivity is 1000Ω/m (b) concrete resistivity is 800Ω/m

(c) concrete resistivity is 600Ω/m (d) concrete resistivity is 400Ω/m

Fig. 11. Corrosion current density cloud picture of track bed rebar under different concrete resistivity when bilateral input.

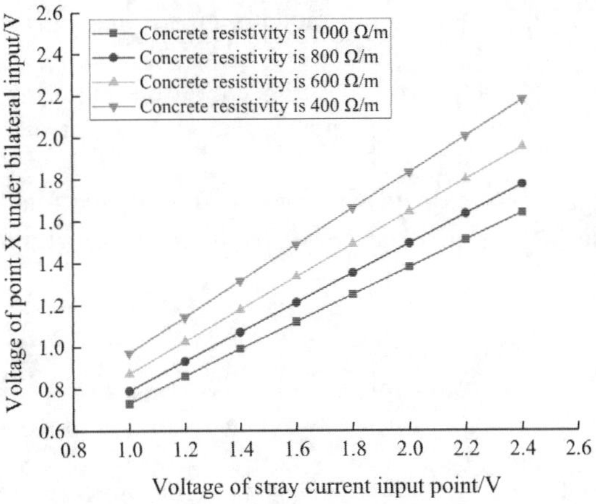

Fig. 12. Relation curve between point X voltage and input point voltage under bilateral input

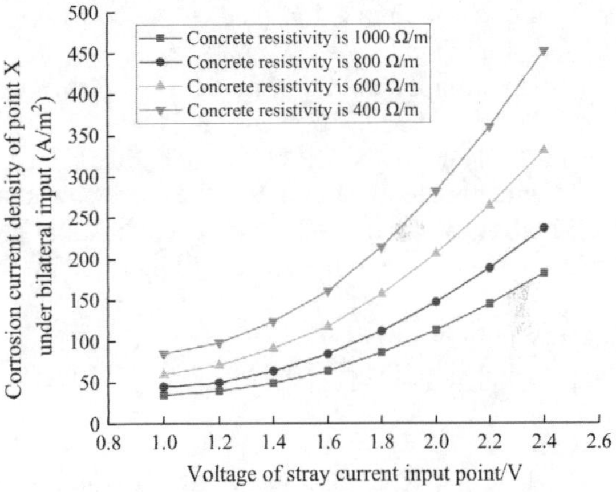

Fig. 13. Relation curve between corrosion current density of point X and input point voltage under bilateral input.

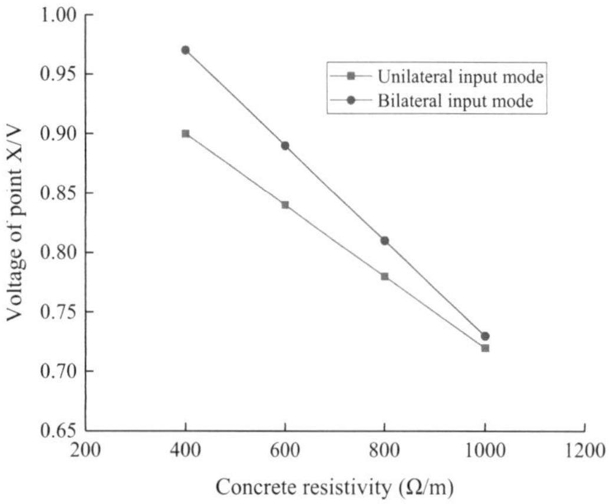

Fig. 14. Relation curve between concrete resistivity and voltage under different input mode

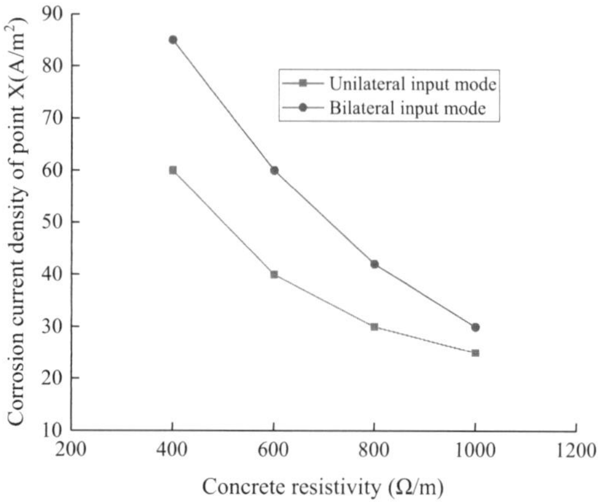

Fig. 15. Relation curve between corrosion current density and concrete resistivity under different input mode

5. Conclusion

In this paper, the effects of stray current input mode and concrete resistivity on the corrosion rate of track bed rebar are studied. The result mainly includes:

(1) The corrosion rate of track bed rebar decreases nonlinearly with the increase of concrete resistivity, and the reduction gradient of corrosion rate in bilateral input mode is greater than that in unilateral input mode.

(2) Under unilateral input mode, the corrosion area of track bed rebar concentrate near the input point when the concrete resistance is 1000 Ω/m, with the decrease of concrete resistance, the corrosion area gradually cover the left half of track bed rebar.

(3) Under bilateral input mode, the corrosion area gradually expands from top to bottom, and finally cover both ends of track bed rebar, when the input point voltage increases by 40%, the corrosion current density in the middle of track bed will increase by about 61%.

Acknowledgements

The present study is sponsored by China Postdoctoral Science Foundation (Grant No. 2022M722958).

References

[1] Jin H, Tian Q R, Li Z, et al. Ability of vibration control using rubberized concrete for tunnel invert-filling[J]. Construction and Building Materials, 2022, 317:125932.

[2] Jin H, Yu S. Study on corrosion-induced cracks for the concrete with transverse cracks using an improved CDM-XFEM[J]. Construction and Building Materials, 2022, 318:126173.

[3] Bertolini L, Carsana M, Pedeferri P. Corrosion behaviour of steel in concrete in the presence of stray current[J]. Corrosion Science, 2007, 49(3):1056-1068.

[4] Dolara A, Foiadelli F, Leva S. Stray current effects mitigation in subway tunnels[J]. IEEE Transactions on Power Delivery[J], 2012, 27(4):2304-2311.

[5] Lei M, Peng L, Shi C. An experimental study on durability of shield segments under load and chloride environment coupling effect[J]. Tunnelling and Underground Space Technology, 2014, 42:15–24.

[6] Gao W, Chen X, Chen D. Genetic programming approach for predicting service life of tunnel structures subject to chloride-induced corrosion[J]. Journal of Advanced Research, 2019, 20:141-152.

[7] Li Z, Jin H, Yu S. Investigation of corrosion rate and rust expansion form of segment reinforcement for shield tunnel by combined action of soil loading, chloride ion and stray Current[J]. Sustainability, 2021, 13(6):3444.

[8] Cheung M, Cao C. Application of cathodic protection for controlling macrocell corrosion in chloride contaminated RC structures[J]. Construction and Building Materials, 2013, 45:199-207.
[9] Kim C, Kim J. Numerical analysis of localized steel corrosion in concrete[J]. Construction and Building Materials, 2008, 22:1129-1136.
[10] Yu S, Jin H. Modeling of the corrosion-induced crack in concrete contained transverse crack subject to chloride ion penetration[J]. Construction and Building Materials, 2020, 258:119645.
[11] Xia J, Li T, Fang J, et al. Numerical simulation of steel corrosion in chloride contaminated concrete[J]. Construction and Building Materials, 2019, 228: 116745.
[12] Jin H, Tian Q R, Li Z. Crack development of rebar rust in rubberized concrete using mesoscale model[J]. Construction and Building Materials, 2022, 321:126409.
[13] Cao C, Cheung M, Chan B. Modelling of interaction between corrosion-induced concrete cover crack and steel corrosion rate[J]. Corrosion Science, 2013, 69:97-109.

AUTHOR INDEX